Introduction to Materials Management Casebook

Revised Edition

J. R. Tony Arnold, P.E., CFPIM, CIRM
Fleming College

Stephen N. Chapman, PhD, CFPIM
North Carolina State University

Lloyd M. Clive, P.E. CFPIM
Fleming College

Upper Saddle River, New Jersey
Columbus, Ohio

Editor in Chief: Stephen Helba
Executive Editor: Debbie Yarnell
Editorial Assistant: Jonathan Tenthoff
Production Editor: Louise N. Sette
Design Coordinator: Diane Ernsberger
Cover Designer: Ali Mohrman
Production Manager: Brian Fox
Marketing Manager: Jimmy Stephens

This book was set in Dutch 801 by Carlisle Communications, Ltd. It was printed and bound by R.R. Donnelley & Sons Company. The cover was printed by Phoenix Color Corp.

Pearson Education Ltd.
Pearson Education Singapore Pte. Ltd.
Pearson Education Canada, Ltd.
Pearson Education—Japan
Pearson Education Australia Pty. Limited
Pearson Education North Asia Ltd.
Pearson Educación de Mexico, S.A. de C.V.
Pearson Education Malaysia Pte. Ltd.

10 9 8 7 6 5 4 3 2
ISBN 0-13-114848-6

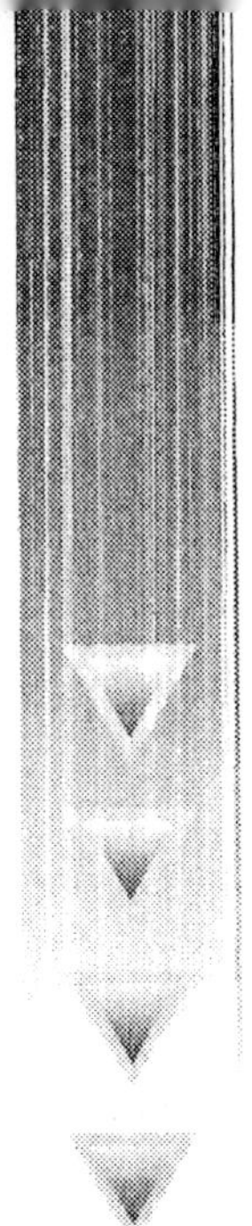

PREFACE

This casebook is intended to be used as a companion to the textbook *Introduction to Materials Management.* However, its usefulness is not limited to this particular text. The cases are designed as examples of the problems that production and inventory management professionals face throughout their working lives.

Students of these cases are typically undergraduates in an introductory course, but the cases also could be used in an introductory course in the master's program, especially with supplemental questions.

This casebook is written to take the student beyond the problems included in *Introduction to Materials Management.* Most chapters have cases, and some cases bridge more than one chapter. Also, the cases vary in level of difficulty, with the more challenging cases requiring the student to think about the management issues involved in their decisions on the job. In all cases, the student has to think beyond the pure "formula application" problem typically found in many textbooks. Generally, the cases will force students to apply multiple concepts and to consider the solution in the context of the situation as opposed to merely finding an answer.

Each of us brings to this *Introduction to Materials Management Casebook* extensive experience in manufacturing and many years of full-time teaching. We welcome your feedback on the content in this text.

We would also like to thank the reviewers of this text: Doug Kopscik, Greenville Technical College; Daniel C. Steele, University of South Carolina; Mehran Hojati, University of Saskatchewan; and Dan Ost, Fox Valley Technical College.

J. R. Tony Arnold,
Professor Emeritus
Fleming College
Peterborough, Ontario
tarnold-15@home.com

Stephen N. Chapman, Ph.D, Associate Professor
Department of Business Management
College of Management
North Carolina State University
stephen_chapman@ncsu.edu

Lloyd M. Clive, Professor, Business Administration
Fleming College
Peterborough, Ontario
lclive@flemingc.on.ca

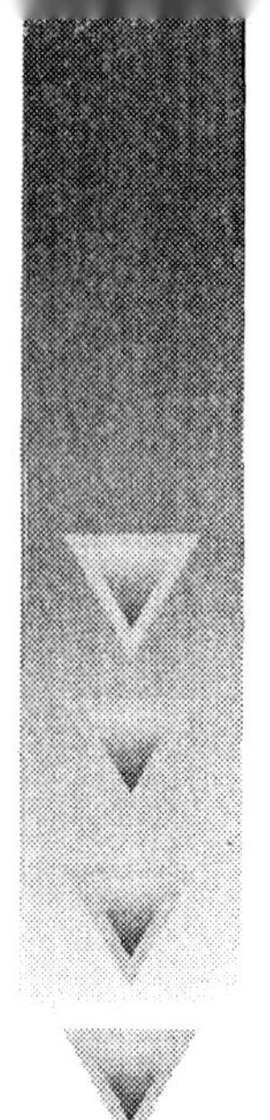

TABLE OF CONTENTS

The following Table of Contents provides a topic for each case. For the convenience of those using *Introduction to Materials Management*, 4th Edition, by Arnold and Chapman, the related textbook chapter is also listed.

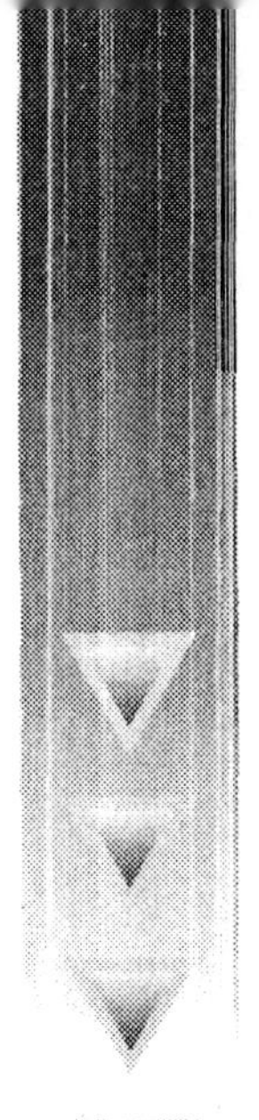

CASE 1

COSTMART WAREHOUSE

Amy Gordon could not have been more pleased when she was first appointed as the new inventory management supervisor for the CostMart regional warehouse. She had previously worked part-time as a clerk in the local CostMart Department Store while she finished her university degree. After she got the degree, she was named as the section head in charge of roughly one-fourth of the store. Now, a year later, she started to wonder about that old adage, "Be careful what you ask for—you just might get it."

BACKGROUND

One constant problem Amy had complained about when she was head clerk was the difficulties she had with the warehouse replenishing supplies for her areas of responsibility. She was sure the problem was not hers. The store used point-of-sale terminals, in which the cash register doubled as a computer, instantly recognizing inventory movement. She also realized that shoplifting and other forms of loss were a constant problem in retail stores, so she instructed all her clerks to spot count inventory in their areas of responsibility whenever there was a "lull" in store traffic. The store computer had a built-in program to suggest replenishment orders when the stock reduced to a certain quantity. Amy had learned, of course, that these were only suggestions, since she knew that some items were "faddish" and would have to be ordered sooner or not reordered at all depending on how the fad was progressing. Some items were seasonal in nature, which needed to be accommodated, and she was also aware when an item would go on sale or have a special promotional campaign. These were announced well in advance during the monthly managerial meetings, and she had good estimates as to the projected impact on demand.

It was because she was so effective at managing the inventory in her area that she was so vocal about the problems at the warehouse. It seemed that almost everything she ordered for replenishment from the warehouse was a problem. Some items were late, occasionally by as many as six weeks. Other items were replenished in quantities far larger or smaller than what was ordered, even if they were occasionally delivered on time. It finally seemed to her that every warehouse delivery was a random event instead of the accurate filling of her orders. Her complaints to general management stemmed from the impact of the warehouse problems. Customers in her area were complaining more often and louder as stockouts of various items became a pattern. Several customers had vowed to never again shop at CostMart because of their frustration. One customer even physically dragged Amy over to the sign above the entrance to the store—the one that proclaims "CostMart—Where Customer Service Is In Charge"—and suggested that she could be sued for false advertising.

In other cases, the quantity delivered was two to three times the amount she ordered. She would often have to hold special "unannounced sales" to avoid being burdened with the excessive inventory, especially since one of her performance metrics was inventory dollars. Of course, one of the major performance metrics was profitability, and both the stockouts and unannounced sales impacted that adversely. Finally, after one particularly frustrating day, she told the general manager, "Maybe you should put me in charge of the inventory over at the warehouse. I can control my own area here—I bet I could put that place back in shape pretty fast!" Two weeks later, she was notified she was "promoted" to inventory management supervisor for the warehouse.

The Current Situation

One of the first issues Amy faced was some not-so-subtle resentment from the warehouse general supervisor, Henry "Hank" Anderson. Hank had been a supervisor for over ten years, having worked his way up from an entry-level handler position. The inventory supervisor position had been created specifically for Amy—Hank had previously had responsibility for the inventory. Their mutual boss had explained to Hank that the reduction in overall responsibility was not a demotion, in that growth in the warehouse made splitting the responsibilities a necessity. While Hank outwardly acknowledged the explanation, everyone knew that in reality he felt the change was a "slap in the face." That would normally be enough to cause some potential resentment, but in addition, as Hank expressed in the lunchroom one day, "It's not enough that they take some of my job away, but then look who they give it to—a young, inexperienced college kid, and a female at that! Everyone knows you can't learn how to run a warehouse in some stupid college classroom—you have to live it and breathe it to really understand it."

Amy knew that the Hank situation was one she would have to work on, but in the meantime she had to understand how things were run, and specifically why the warehouse was causing all the problems she experienced at the store. Her first stop was to talk to Jane Dawson, who was responsible for processing orders from the store. Jane explained the situation from her perspective,

"I realize how much it must have bothered you to see how your store requests were processed here, but it frustrates me too. I tried to group orders to prioritize due dates and still have a full truckload to send to the store, but I was constantly having problems thrown back at me. Sometimes I was told the warehouse couldn't find the inventory. Other times I was told that the quantity you ordered was less than a full box, and they couldn't (or wouldn't) split the box up, so they were sending the full box. Then they would find something they couldn't find when it was ordered a long time ago, so now that they found it they were sending it. That order would, of course, take up so much room in the truck that something else had to be left behind to be shipped later. Those problems, in combination with true inventory shortages from supplier missed shipments always seems to put us behind and never to be able to ship what we are supposed to. None of this seemed to bother Hank too much. Mayby you can do something to change the situation."

Amy's concern with what Jane told her was increased when she asked Jane if she knew the accuracy of their inventory records and was told that she wasn't sure, but the records were probably no more than 50% accurate. How can that be?, Amy asked herself. She knew they had recently installed a new computer system to handle the inventory, they did cycle counting on a regular basis, and they used a "home base" storage system, where each stock-keeping unit (SKU) had its own designated space in the warehouse racks. She realized she needed to talk to one of the workers. She decided on Carl Carson, who had been with the company for about five years and had a reputation of being a dedicated and effective worker. Amy told Carl what she already knew and asked him if he could provide any additional information.

According to Carl, "What Jane told you is true, but what she didn't tell you is that a lot of it is her fault. If she would only give us some advanced warning about what she wants to send for the next shipment we could probably do a better job of finding the material and staging it. What happens, though, is that she gives us this shipment list out of the blue and expects us to find it all and get it ready in very little time. For one thing, she doesn't understand that it's very impractical to break boxes apart in order to ship just the quantity she wants. We don't have a good way to package the partial box, and an open box increases the chance for the remaining goods to be damaged or get dirty. Even if we had a way to partial package, the time it would take would increase the chance we wouldn't make the shipment on time.

"Then there's the problem of finding material. When supplier shipments come in, they are often for more goods of a given SKU than we have room for on the rack. We put the rest in an overflow area, but it's really hard to keep track of. Even if we locate it in the system correctly, someone will soon move it to get to something behind it. They will usually forget to record the move in the heat of getting a shipment ready. Since the cycle counts don't find it in the designated rack, the cycle counters adjust the count so the system doesn't even know it exists anymore. You might think we should expand the space in the rack to hold the maximum amount of each SKU, but we would need a warehouse at least double this size to do that—and there's no way management would approve that. I guess the only good thing about the situation is that when we do find some lost material that was requested earlier, we ship it to make up for not shipping it earlier."

Amy was beginning to feel a tightening in her stomach as she realized the extent of the problem here. She almost had to force herself to talk to Crista Chavez, who worked for the purchasing department and was responsible for warehouse ordering. Crista was also considered to be experienced, capable, and dedicated to doing a good job for the company. Crista added the following perspectives,

"We have good suppliers, but they're not miracle workers. Since we beat them up so badly on price most of the time, I can understand why they're not interested in doing more than they already are. The problem is we can't seem to get our own 'house in order' to give them a good idea what we need and when we really need it. To do that, we would need to know what the warehouse needs and when, and also the existing inventory of the item. We seem to have no idea what we need, and the inventory records are a joke. I spend most of my day changing order dates, order quantities, or expediting orders to fill a shortage—and often the shortage isn't really a shortage at all. Our only hope has been to order early and increase our order quantities to ensure we have enough safety stock to cover the inventory accuracy problems. I've complained to Hank several times, but all he says is that it's my job to pull the suppliers in line, that the problem is obviously theirs."

At least by this point Amy had a better perspective about the problems. Unfortunately, it was now up to her to fix them. She wished she had never opened her mouth to complain about the problems. Too late for that—she now had to develop a strategy to deal with what she had been handed.

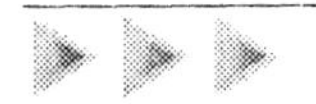

CASE ANALYSIS

1. Structure what you think the problems are. Be sure to separate the problems from the symptoms.
2. Assume Amy needs to build a data-based case to convince her boss and start to "win over" Hank. What data should she gather to help her build the case?
3. Develop a model of how you think the warehouse should work in this environment.
4. Develop a time-phased plan to move from the present situation to the model you developed in question 3.

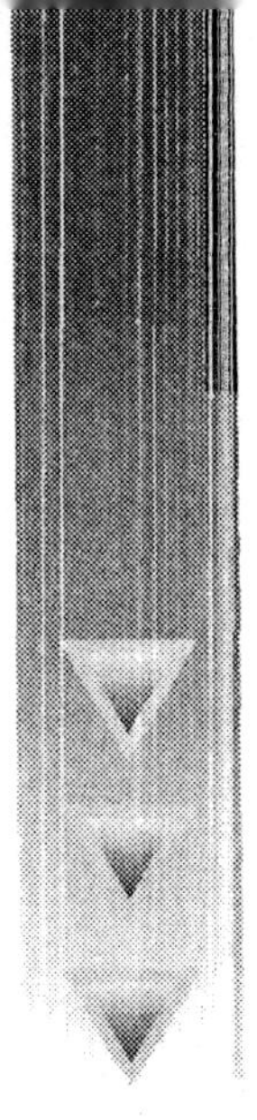

CASE 2

JOHNSTON PRODUCTS

No matter how many times Justin Wang, the Master Scheduler for Johnston Products, tried, he couldn't seem to get it through people's minds. They kept trying to "front-load" the production schedule, and the problem appeared to be getting worse.

By "front loading" Justin meant that production supervisors would attempt to catch up with production they failed to make the previous week. It seemed to happen every week, and the only way Justin could get things back to a realistic position was to completely reconstruct the entire master schedule—usually about every three weeks.

Last month could serve as an example. The first week of the month Justin had scheduled production equal to 320 standard hours in the assembly area. The assembly area managed to complete only 291 hours that week because of some equipment maintenance and a few unexpected part shortages. The assembly supervisor then had the workers complete the remaining 29 hours from week 1 at the start of week 2. Since week 2 already had 330 standard hours scheduled, the additional 29 hours really put them in a position of attempting to complete 359 hours. The workers actually completed 302 hours in week 2, leaving 57 hours to front-load into week 3, and so forth. Usually by the time Justin came to his three-week review of the master schedule, it was not uncommon for the assembly area to be more than 100 standard hours behind schedule.

Clearly something needed to be done. Justin decided to review some of the areas that could be causing the problem:

1. Job standards—While it had been at least four years since any job standards had been reviewed or changed, Jason felt the standards could not be the problem—quite the opposite. His operations course had taught him about the concept of the learning curve, implying that if anything the standards

should be too large, allowing the average worker to complete even more production than implied by the job standard.

2. Utilization—The general manager was very insistent on high utilization of the area. He felt that it would help control costs, and consequently used utilization as a major performance measure for the assembly area. The problem was that customer service was also extremely important. With the problems Justin was having with the master schedule, it was difficult to promise order delivery accurately, and equally difficult to deliver the product on time once the order promise was made.
3. The workers—In an effort to control costs, the hourly wage for the workers was not very high. This caused a turnover in the workforce of almost 70% per year. In spite of this, the facility was located in an area where replacement workers were fairly easy to hire. They were assigned to the production area after they had a minimum of one week's worth of training on the equipment. In the meantime, the company filled vacant positions with temporary workers brought in by a local temporary employment service.
4. Engineering changes—The design of virtually all the products was changing, with the average product changing some aspect of the design about every two months. Usually this resulted in an improvement to the product, however, so Justin quickly dismissed the changes as a problem. There were also some engineering changes on the equipment, but in general little in the way of process change had been made. The setup time for a batch of a specific design had remained at about 15 minutes. That forced a batch size of about from 50 to 300 units, depending on the design. The equipment was getting rather old, however, forcing regular maintenance as well as causing an occasional breakdown. Each piece of equipment generally required about three hours of maintenance per week.

Since the computer had done most of his calculations in the past, Justin decided to check to see if the computer was the source of the problem. He gathered information to conduct a manual calculation on a week when there were eight people assigned to the assembly area (one person for each of eight machines) for one shift per day. With no overtime, that would allow 320 hours of production.

Product	Batch Size	Standard Assembly Time (minutes per item)
A174	50	17
G820	100	9
H221	50	19.5
B327	200	11.7
C803	100	21.2
P932	300	14.1
F732	200	15.8
J513	150	17.3
L683	150	12.8

CASE ANALYSIS

1. With this information, Justin calculated the total standard time required to be within the 320 hours available. Is he correct? Calculate the time required and check the accuracy of his calculation.
2. List the areas you think are causing trouble in this facility.
3. Develop a plan to deal with the situation and try to get the production schedule back under control under the constraints listed.

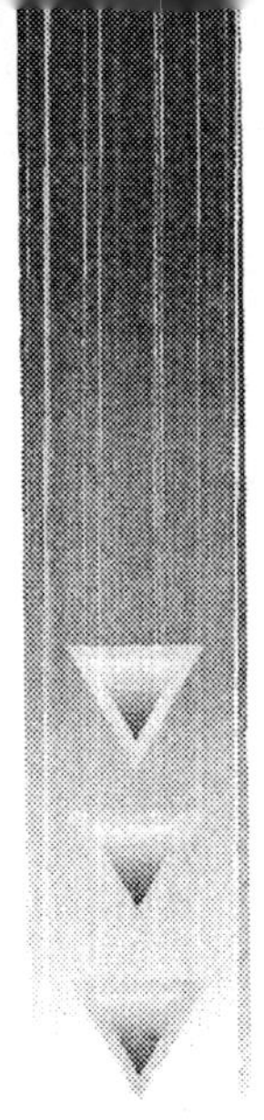

CASE 3

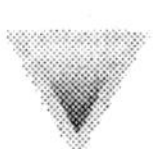

KATHY'S KREAMY KREATIONS

About seven years ago, Kathy Horton took a great interest in dessert specialties. It started because she had a slight weight problem, but loved desserts, especially sweets. To deal with the problem, she decided to learn as much as possible about desserts so she could appease her sweet tooth without buying a whole new wardrobe of larger sizes.

As her knowledge grew, she started to make several of the desserts—at first for herself, but soon her experiments were in great demand among her friends and family. They became so popular and she received so much encouragement that she decided to go into the dessert business on evenings and weekends.

HOW THE BUSINESS DEVELOPED

At first Kathy worked out of her home, generating sales through word of mouth and a small advertisement in the yellow pages. Most of her initial sales were for special occasions, such as birthday parties and anniversaries. Soon she found that she was making the majority of sales to other businesses, such as restaurants and specialty grocery stores. As sales increased substantially, she outgrew the ability to use her own kitchen. She also thought that if she could increase sales just a little more, she could quit her regular job and devote full time to the dessert business.

Things didn't work out quite as Kathy planned. In order to find a kitchen suitable for her needs, she leased a space that was previously a small restaurant. It was the only place she could find that came even close to having the type of space and equipment she needed. Even though not ideal, she could not afford to build and equip the "perfect site" for the business. Not only was the lease payment more than she anticipated, but there was also a large space she had no real use for (what used to be the restaurant eating space). She decided to use the space to generate extra revenue (to cover the lease expense) by making extra desserts and selling them on a piece-by-piece basis to walk-in customers. She would also try to sell some desserts over-the-counter rather than exclusively on a make-to-order basis. This forced her to add two salespeople for the "front end" of the business.

Kathy had hoped that hiring the salespeople could free her from the walk-in business, but it was obvious that it could not. Since the walk-in business was fairly small, especially at first, she could only afford to pay minimum wage. The people she got to fill the positions were, as might be expected, unskilled and lacked any knowledge about the desserts that they were selling. Kathy found she was spending time to both train and supervise, and even the training didn't stop. The minimum wage was causing a high employee turnover, so she found herself constantly having to rehire and train new people. Even with the minimum wage, the revenue seldom covered costs—but at least she felt she was helping to cover the overhead for that previously unused space.

The changes were not over for Kathy. She now found that the total demand from her regular customers (make-to-order) and her new walk-in customers did not allow her to both manage the business and make the desserts especially with a large amount of time taken up with training, hiring, and supervision. She decided to hire a local homemaker (Mary) part-time to help make the desserts and also hired a recent graduate from a master's program in management to keep the books and manage the "front end" part of the business.

The extra expense from these hires again put her badly in the red. To increase revenue and to attract more customers (and hopefully become profitable), she decided to add meals to her menu for the walk-in customers. Soon she was offering a fairly extensive luncheon and dinner menu, although her specialty and reputation were still focused on her dessert creations.

The expanded menu and associated hours caused two additional problems. The first was the need to hire even more people (essentially representing two complete shifts of restaurant workers—including cooks and servers), which once again significantly increased her break-even revenue requirements. The second problem was that the restaurant meal business required more and more of her kitchen space. She often found her dessert production requirements and her restaurant requirements in competition for space. To solve the problem, she would often show up in the kitchen at 3 A.M. in order to have room to work. Unfortunately, she also felt obliged to be around the restaurant for most of the day. After all, it was her business and she felt she needed to keep a firm watch on it.

Another situation that she had never considered was beginning to be an issue. Once she started the walk-in and restaurant business she was in fact becoming a competitor to some of her best customers—namely, some of the other restaurants in town. They were beginning to complain, charging that she perhaps had used them to build her own reputation and now was undercutting them. She tried to explain that she was only doing the walk-in and restaurant business to help with expenses, but she was sure that some (and perhaps all) of those customers did not really believe that story even though it was the truth. From their perspective the issue was plain—someone who had once been a loyal and high-quality supplier had expanded into their own business and had, therefore, become a competitor.

One day while taking a rare day off, she reflected on her situation. After almost five years, she was getting "burned out," was no longer enjoying her work, was putting in 15-hour days at least six days a week, and the business was barely profitable. She felt she had been pulled away from the original focus of the business (and the focus that she enjoyed), and now found almost no time to be creative and attempt new recipes. Her core customers were becoming unhappy, and without new and better creations, how could she hope to keep them? Even with new creations, how could she convince them she really didn't view herself as their competition? How had she gotten herself into this mess, and more importantly, how could she get out of it?

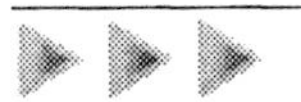

Possible Courses of Action

None of the alternatives Kathy considered appeared attractive. They included the following:

1. She could get rid of the restaurant and go back to working out of her kitchen. To do so would not only force her to drop some good customers, but would also place a severe limit on her earning potential. Even though this alternative would allow her to move back to working the way she enjoyed, the only way she did it before was to work at another job to make enough to live on. She felt that taking this move was not only a step backward but would leave her with no way to progress again.
2. She could concentrate on just the restaurant. She was highly concerned that most of her restaurant customers were loyal because of the desserts. Without her working on those creations, would she become just "another restaurant" without the distinctive draw that her desserts generated? After all, she had purposely tried to limit the restaurant menu for meals to keep that part of the business as simple (and hopefully as cost-efficient) as possible. Also she was concerned that her real love and interest were in the desserts, and the restaurant was only a sideline to help pay the bills in the first place. Could she really make it in the competitive restaurant business, especially since her heart wasn't really in it?
3. She could stay in the building, but only work on desserts. This would imply wasted space, but would she be able to generate enough business in just desserts to pay for the space? This was especially bothersome since her regular dessert customer base had grown to the point where she would need to keep Mary on the payroll. Business was rather erratic (focused on holiday time primarily), so cash flow was an issue. Could she afford to continue to pay Mary during the times when she really didn't need her? Mary was proving herself to be very good, and Kathy felt that in her type of business quality was of prime importance.

4. She could give up the building and build a special kitchen. There were no "pure kitchens" available for lease or purchase—she had already determined that in her original search. She could build a space or buy a vacant space and furnish it, but the loan payments were potentially more risky than her current lease payments because of the timing issues.

None of her options seemed appealing. Even the thought of giving up and working for someone else again bothered her a great deal. She now had the entrepreneurial desire, and combined with her knowledge of desserts (and to some degree business) she knew there must be a way.

CASE ANALYSIS

1. What, if anything, did Kathy do "wrong" to get herself into this situation? What should she have done differently? Outline with as much structure as possible using the concepts about service operations and service designs developed in class.
2. What analysis should she do to help her make a decision? Design a complete program of analysis, describing what data to collect, how to organize it, and what to do with it.
3. Given the limited data in the case, what would you suggest doing and why? Develop a comprehensive plan for Kathy's business and provide sound operational design justifications for your suggestions, again drawing on issues discussed in class.

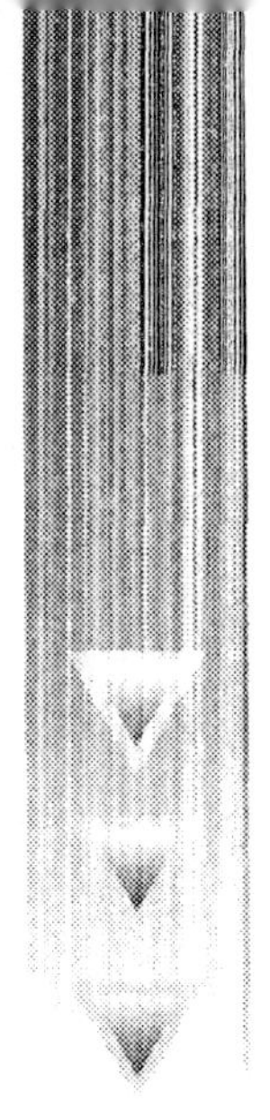

CASE 4

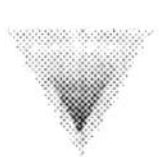

LET'S PARTY!

The words above are still echoing in your head as you leave your Principles of Buying class. Again you ask yourself, "Why did I ever let myself run for class President?" Most of the people in the class were good, level-headed individuals who enjoyed a good time and you enjoyed working with them. But a small group from your class, who were known on campus as The Rowdies, often bullied their way on decisions affecting class activities. The decision to have a year-end party was right up their alley and class had ended with a chanting session of "Let's Party." It sounded like a wrestling match to you. Fortunately, your professor had left the room early to let you discuss with the class the idea of some kind of year-end get-together.

The Rowdies had immediately suggested the Goat's Ear, a local hangout with not much to offer but cheap drinks. The rest of your classmates had put forth some other suggestions, but no consensus on a location could be reached between the members of your executive committee or the rest of the class. If you went to the Goat's Ear most of the sane people in your class wouldn't attend, and even when you suggested more conventional locations people couldn't agree because of factors such as the type of music played.

Since there were only two weeks left until the end of regular classes you felt that you had to make arrangements in a hurry. It wasn't difficult to identify the most popular possible locations, but getting agreement from this group was going to be difficult.

One of your recent lectures was on supplier selection and your professor had demonstrated the technique called the ranking or weighted-point method. It seemed simple enough in the lecture and you had almost embarrassed yourself by asking the question, "Why not just pick the least expensive supplier?" The thought occurred to you that there just might be some solution to your current problem in the professor's response, "One of the hardest things to do in any group, whether a business or a social club, is to get consensus on even the simplest choices."

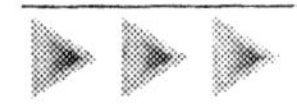

CASE ANALYSIS

For this exercise, put yourself in the position of the person above and complete one of the two following exercises:

EXERCISE 1

1. Perform a supplier rating analysis for the above situation. You should include at least ten factors and four possible locations.
2. Make the selection as indicated by the analysis.
3. Discuss why the analysis made the selection as in step 2 and whether you would change any of the criteria or weights.

EXERCISE 2

1. Prepare a transparency to be used in class to make a selection for a year-end get-together.
2. Lead a discussion to determine at least four possible locations and ten factors.
3. Have the class agree on weighting factors for each criteria.
4. Perform the calculations and make the selection.
5. Discuss with the class why the analysis made the selection as in step 2 and whether you would change any of the criteria or weights.
6. Ask the class whether they feel more in agreement with the decision after going through this process.

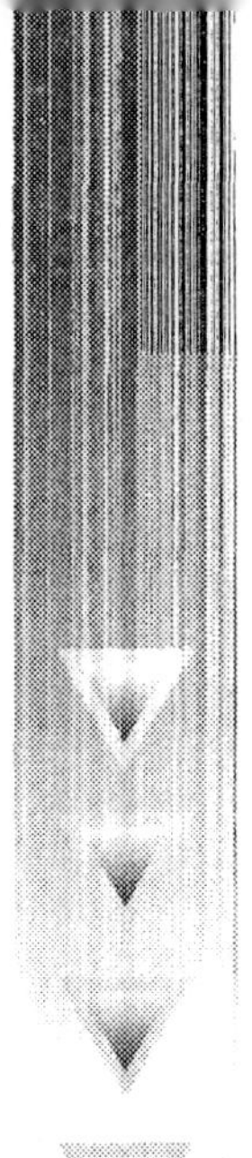

CASE 5

NORTHCUTT BIKES I: THE PRODUCTION PLANNING PROBLEM

Near the end of December 1997 Jan Northcutt, CEO of Northcutt Bikes, entered the conference room for another planning session for 1998 production. Other key participants in the meeting included Jenny Hansen, Director of Marketing, Tim Weller, Director of Production, and Larry Ball, Manager of Inventory and Distribution.

Jenny started the meeting by reviewing the latest sales projections and situation. "I've been discussing the latest situation with our sales representatives, and they strongly feel we lost a lot of potential sales last year because we were not as responsive or flexible to special orders as we were in the past. If we can all pull together, the sales staff feels that in 1998 we can easily sell more than 17,000 of the standard 26-inch 10-speed street bike alone, and they have similar optimism for other models as well."

At that point Tim jumped up and interrupted by saying, "Come on, Jenny, you've got to get realistic. Less than a month ago you presented a sales forecast of 14,000 units of that model for next year. Now you've raised that projection by over 20%! How do you expect us to ever be able to effectively plan production when you do that to us?"

Jenny felt she had to respond. "Do you want us to tell you what we can really sell or only what we have available to sell given your inability to be responsive to our customers? We could have sold more than 15,000 of that model even last year if your production capabilities hadn't held us back so badly. We built this company on customer service and now you're slowly destroying our competitive advantage with your production problems! You always seem to be slow on special orders and if you would produce more inventory for us to sell we wouldn't have the response problem."

At this point Larry had to break in. "Forget about trying to add inventory. It's not only costing us a lot of money to hold on to what we have, but we're basically out of room in the warehouse. I'd have to lease more space, and that won't come cheaply! It would do us very little good to be much more responsive if we also became much more expensive. Besides, the inventory always seems to have the wrong options. I would guess that most of the inventory in the warehouse ends up going back to the production floor for option rework before shipment."

Jenny again responded, "I don't know why the production problems happen, but I do know that several of our best customers have threatened to drop our lines if we don't do a better job this year of providing order service. Back orders are not healthy for a company founded and built on service and flexibility!"

Jan could take no more. "I've heard enough of this bickering. We're supposed to be a cohesive management team with a coordinated goal of maximizing service to our customers while maximizing profit. Instead this meeting sounds like a petty grade school argument. Jenny, I want you to get your best information to Tim as soon as possible. Tim, take the information and develop your best production plan possible. Be sure to include costs of inventory, hiring, firing, and overtime. You may need to get some inventory cost data from Larry first. If we have to add a second shift, I'll need the best estimate as to when we need it and we'll have to be sure it isn't more expensive than adding more leased warehouse space. The one thing I won't tolerate this year is the kind of back order condition that started to sour our relationship with our customers.

"Try it first with the 26-inch bike line and if it looks good we can expand the process to our other lines. The 26-inch line is best to use because it is both representative of other lines and is also relatively self-contained.

"In any case, we've got to move fast on this. We have less than two weeks to go before 1998 sales start, and we've got to be ready!"

Description of the 26-inch Line

There are four basic models produced on the 26-inch heavy-duty line. They include the standard 10-speed street bike, the deluxe 10-speed street bike, the standard heavy-duty, and the deluxe heavy-duty. While many of the parts are the same for all models, the frame differences and the gearing differences do imply some changeover costs when moving from production of one model to another. For the purposes of production planning, however, the changeover costs are assumed negligible.

Labor Costs

While Northcutt Bikes purchases less than 50% of the value of the final product, the bikes are produced on a fairly integrated production area. The workers are mostly semiskilled, with the average beginning worker being paid $10.50 per hour plus $3.20 in benefits. The wage climbs to a maximum of $16.80 for an experienced worker, but the average for all workers over the last couple of years has been $12.70 plus $3.50 in benefits. They decided to assume the average wage in making calculations. While it varies somewhat on options, on the average each bike utilizes 10 worker-hours of these direct laborers to complete.

To this point, the company has not utilized a second shift, even though they can only use a maximum of 150 direct workers on any given shift. They can use overtime at 150% of the base pay, with no change in benefit cost, but are limited by agreement to only eight hours per worker per week of overtime. If they do use a second shift, the base pay would increase by 20% due to a shift premium. Benefit cost would be the same for a second-shift worker.

With training and on-the-job learning, the average new employee takes about three weeks to reach a full productivity level. Even so, for production planning purposes they decided to assume full productivity at the moment of hire. On average, the cost of hiring and training a new employee is about $1100. If an employee is ever laid off, the company estimates an average cost of $1800 to cover unemployment and other expenses. There are currently 110 of these direct labor employees. This figure for total direct labor employees is lower than the peak in the summer, but was basically achieved by attrition.

Additional costs are incurred for indirect labor and management. The company estimates that for every 50 direct labor employees, they need two supervisors, one clerk, and two maintenance people. The total cost for those five people is estimated at $200,000 per year, including benefits. Because they have not laid off any of these indirect people (or direct employees) since the summer peak demand, they currently have six supervisors, three clerks, and six maintenance people. While Jan's past policy has been to try to avoid layoffs, she has stated that she will definitely consider layoffs if they improve the cost picture significantly.

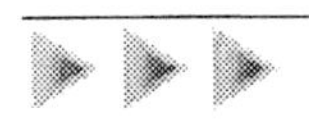

Inventory Costs

Larry presented Tim with the summary inventory data. He currently has 1100 standard 10-speed street bikes, 550 deluxe 10-speed street bikes, 270 standard heavy-duty, and 220 deluxe heavy-duty bikes in finished goods inventory. He estimates he only has room for 4500 total bikes of this type in the warehouse. The cost to keep one bike for one month in the warehouse is currently about $9 per month, including cost of capital, warehouse costs, and average rework costs. He has checked with local public warehouses and finds the cheapest cost runs about $5 per bike per month, not including the average $7 per bike per month of capital cost and rework cost incurred when they must withdraw a bike from finished goods to rework it for a specific customer order.

The following schedule gives the expected working days for the year:

Month	Weeks	Working Days
January	5	20
February	4	20
March	4	23
April	4	18
May	5	21
June	4	23
July	5	21
August	4	21
September	4	21
October	5	21
November	4	20
December	4	21

MARKETING BUSINESS PLAN

The following table represents the 1998 forecast for each heavy-duty bike model presented to Tim by Jenny:

Month	Standard 10-Speed Street Bike	Deluxe 10-Speed Street Bike	Standard Heavy-Duty	Deluxe Heavy-Duty
January	950	620	250	270
February	1140	660	350	300
March	1280	770	410	340
April	1490	840	430	440
May	1610	920	520	470
June	1820	1010	630	540
July	1790	930	560	510
August	1630	850	510	460
September	1490	790	460	400
October	1350	740	400	360
November	1220	700	370	320
December	1130	640	330	290

CASE ANALYSIS

1. Develop at least three unique plans for meeting the sales estimates with low cost as a goal. Be sure to include all relevant costs, including hiring, firing, inventory, overtime, shift premium, and indirect labor. Do not try to optimize a solution, but instead try to develop a sensible heuristic solution. What major problems did you encounter as you developed the data that are worth making Jan aware of?
2. Are there any policy changes you would recommend reviewing? What are they, why are you concerned, and what direction might you suggest they examine?
3. What possible problems do you see as a result of the assumptions made to simplify this process? Do these assumptions suggest additional investigation possibilities for Northcutt?
4. What is your overall recommendation to Jan?

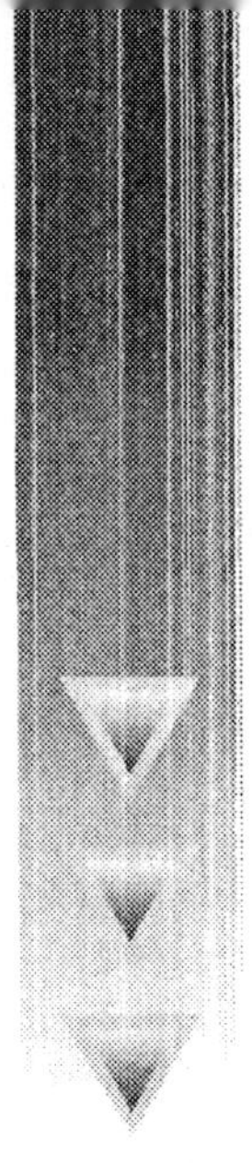

CASE 6

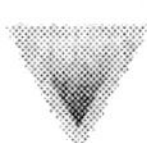

FRAN'S FLOWER SHOP

After Fran graduated with an undergraduate art degree in 1988, she decided to combine her knowledge and love of art with a second love—plants and flowers—toward developing a business. Her intent was to focus on a specialty niche in the flower shop business. She decided to concentrate her efforts on made-to-order special flower arrangements, such as typically found at banquets and weddings. Due to her talent and dedication to doing a good job, she was highly successful, and her business grew to where she now owns Fran's Flower Shop, located in a highly visible and successful strip mall. As with many successful businesses, her success has produced unanticipated problems, some of which are normal "growth pains," but others are relatively unique to the type of business. At a recent meeting with her business adviser, she outlined four of the major issues she faces.

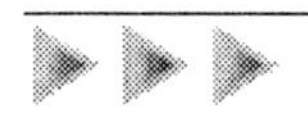

BUSINESS FOCUS

When she moved into her new shop in the mall, she continued to specialize in the make-to-order specialty arrangements, but customers frequently walked into her shop requesting "spot" purchases, including gifts for sick friends and last-minute flower purchases for occasions such as birthdays, anniversaries, Valentine's Day, and so forth. As this business represented an attractive addition to the store revenue, she accommodated it with three large, climate-controlled display cases stocked with ready-to-sell arrangements of various sizes, types, and prices. Even though she did not aggressively pursue this market with advertising, the heavy mall traffic where her store is located and word of mouth caused the walk-in business to steadily grow to where it now represents almost half of her total revenue. This business has brought her numerous headaches, however, due to several characteristics:

1. Even though some days have predictably high demand (e.g., just prior to Valentine's Day, Mother's Day), most of the time she has no idea as to how many customers will come in for spot buys on any given day, nor does she have any idea as to the price range they will look for. Even such variables as the weather and the schedule of local sports teams appear to affect her demand. She knows she needs to manage this demand better, because not having what a customer wants could mean the permanent loss of a good potential customer. On the other side, flowers have a limited shelf life, and having too much of the wrong price range could mean a high spoilage rate. It would not take many totally lost arrangements on a daily basis to represent the difference between profit and loss for that part of the business.
2. Some customers have become irate that her delivery system (a major part of the make-to-order business) will not accommodate the delivery of a $20 ready-to-sell arrangement to a hospital, for example. Angry customers have even asked her how much they need to spend on an arrangement before she will deliver. She has never really thought about an answer to that question and has not known how to reply. Generally she just states that she does not deliver premade flower arrangements. She knows this lack of delivery has cost her some good will, some business, and perhaps even some potential return customers.
3. Related to the point above, several customers have expressed serious dissatisfaction that Fran's Flower Shop is not a member of some national delivery service (such as FTD), so they can have flowers delivered out of town. Fran is afraid such a business will pull her even further from her core business of make-to-order, as those services typically focus on catalogs of set designs. Since those services are also expensive to belong to, she knows she would have to spend a lot more time and effort in that area to make it financially feasible.
4. Another group of customers wants her to extend her open shop hours because they occasionally drop by for flowers on their way home from work and often find her closed for the day or at least not available while she is setting up a flower order in some other location.

Personnel Issues

As her business grew, Fran hired another skilled arranger (Molly) to work with her. The unpredictability of the walk-in demand has caused her to bring "people" issues up as a problem, however. Since walk-in customers demand immediate attention, she and Molly are frequently called to the front of the shop to sell arrangements from the cases. This pulls them away from working on their orders, and while she has only been late on a couple of special orders within the last few weeks, several others were delivered before she was satisfied with their appearance, merely to avoid their being late. This worries her a great deal, as she has worked very hard to obtain a reputation for the quality of her arrangements. She thought about hiring a delivery person, but decided it was important that either she or Molly deliver the orders so that they may put last-minute touches on the arrangement in case of disturbance during the delivery process.

Instead, she opted to hire some part-time unskilled workers for the shop to handle the walk-in sales. This has proved less than satisfactory because of two reasons:

1. The unpredictability of demand has her constantly wondering about what hours and how many hours to schedule the workers for. The extra help adds to cost, and those hours paying someone to stand around while no customers come into the shop make the difference between profit and loss even more sensitive.
2. Customers frequently have questions about the type of flowers in an arrangement, how long they last, and so forth. The unskilled workers she hires often don't know. They will then frequently interrupt either Fran or Molly with the question, and even when they get the answer the customer is often left with a poor impression, since they usually expect more knowledge from a salesperson. The impression is even worse if Fran and Molly are both out servicing orders, since the only answer the customer gets is "I'm not sure." Because she pays only slightly above minimum wage, her turnover is high. This means she is constantly trying to hire and train people, further distracting her from her main business. She knows she could reduce the turnover and hire more knowledgeable people if she paid her help more per hour, but that issue again pushes her closer to the loss column for many of the days the shop is open, so she feels she really can't afford to pay more.

EXPANSION

Several of her regular customers are encouraging her to open another operation on the other side of the city, as well as considering expansion to other cities. They claim several of their friends like her arrangements a great deal, but consider her location too inconvenient from where they live or work. That is typically not a problem for large orders, since she or Molly will typically offer to visit the customer to obtain details for the arrangement. That does take a lot of time, however, so she finds herself more frequently asking the long-distance customer to come to the shop if possible. Many decline to do so, and the order is sometimes lost. While expansion is attractive to her, she worries about control—not only for order servicing, but also for delivery. How can she possibly maintain control of quality and design in two or more locations at once?

SUPPLY

Since her purchases of flowers from the wholesaler have grown, the wholesaler has recommended that Fran make a purchasing contract instead of making spot bulk buys as she now does. This contract will give her significant quantity price discounts, but her delivered quantity has to represent above a certain amount of each type of flower so that the wholesaler can reduce costs due to economies of scale. The quantities she needs to order are reasonable given her average demand, but the fluctuation around that average is large enough to present significant spoilage during certain periods. She wonders if she would be better off in the long run with the purchasing contract.

CASE ANALYSIS

1. What are the key issues in this case? In other words, analyze the case to try to determine the true problems from the symptoms of those problems. How do these issues relate to the issue of process design for services?
2. What type of data would you suggest collecting to both verify that the problem identifications are correct as well as to provide solution approaches and support? How would you organize and use that data?
3. What would you suggest Fran do with her business and why? Provide a comprehensive and integrated plan of action and provide support for your suggestions by referring to service process design issues discussed in class as well as from any other source.
4. Develop an implementation plan for whatever changes you suggest she make. Prioritize the key steps if appropriate.

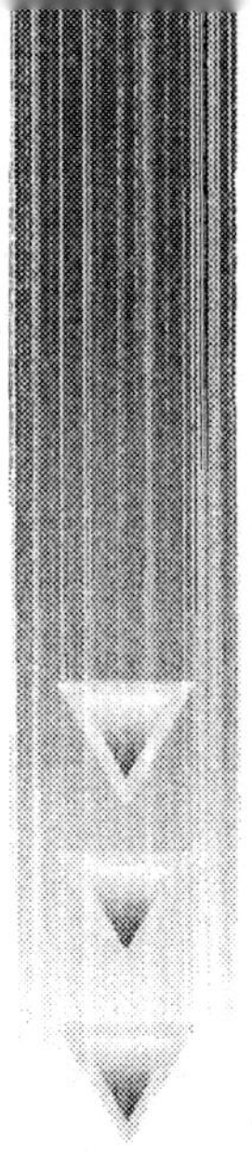
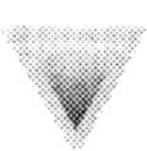

CASE 7

ACME POLYBOB COMPANY

Ken Mack, plant manager for the Acme Polybob Company, was having a heated discussion with Jack Gould, the production and inventory control manager. Ken was getting tired of frantic calls from Ellen Uphouse, the marketing manager, concerning late orders for their Polybob customers and was once again after Jack to solve the problem. Some of the discussion points follow:

Jack: "Look Ken I'm not sure what more we can do. I've reexamined the EOQ values and all the reorder points for all our inventory for all our Polybob models including all component levels and purchased items. I've implemented strict inventory control procedures to ensure our accuracy levels to at least 80%, and I've worked with the production people to make sure we are maximizing both labor efficiency and utilization of our equipment. The real problem is with those salespeople. We no sooner have a production run going nicely and they change the order or add a new one. If they'd only leave us alone for a while and let us catch up with our current late order bank we'd be okay. As it is, everyone is getting tired of order changes expediting and making everything into a crisis. Even our suppliers are losing patience with us. They tend to disbelieve any order we give them until we call them up for a crisis shipment."

Ken: "I find it hard to believe that you really have the EOQ and reorder point values right. If they are, we shouldn't have all these part shortages all the time while our overall inventory is going up in value. I also don't see any way we can shut off the orders coming in. I can imagine the explosion from Ellen if I even suggested such a thing. She'll certainly remind me that our mission statement clearly points out that our number one priority is customer service and refusing orders and order changes certainly doesn't fit as good customer service."

Jack: "Then maybe the approach is to deal with Frank Adams (the chief financial officer). He's the one who is always screaming that we have too much inventory, too much expediting cost, too much premium freight costs from suppliers, and poor efficiency. I've tried to have him authorize more overtime to relieve some of the late order conditions but all he'll say is that we must be making the wrong models. He continually points to the fact that the production hours we are paying for currently are more than enough to make our orders shipped at standard and that condition has held for over a year. He just won't budge on that point. Maybe you can convince him."

Ken: "I'm not sure that's the answer either. I think he has a point and certainly has the numbers to back him up. I'd have a really rough time explaining what we were doing to Roger Marrison (the CEO). There's got to be a better answer. I've heard about a systems approach called material requirements planning or something like that. Why don't you look into that? Take a representative model and see if that approach could help us deal with what appears to be an impossible situation. I'm sure something would work. I know other factories have similar production conditions yet don't seem to have all our problems."

The following is the information about Polybob Model A that Ken suggested as a representative model to use for the analysis:

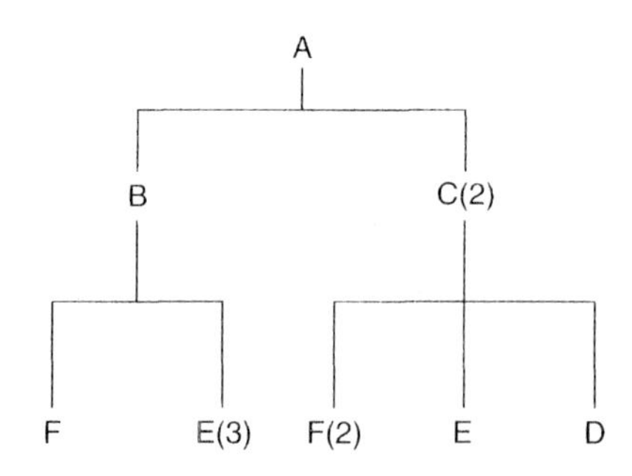

Component	Lot Size	Inventory	Lead Time	Scheduled Receipts	Reorder Point
B	Lot for lot	10	1	None	5
C	150	40	1	None	15
D	200	180	2	None	50
E	400	400	2	None	70
F	500	50	2	500,week 1	80

The following are the master schedule production lots for Model A:

Complete 50 units, week 3
Complete 50 units, week 5
Complete 60 units, week 7
Complete 60 units, week 9
Complete 50 units, week 11

Upon seeing this information, Ken stated, "Look at how regular our production schedule is for this model. The reorder points will more than cover requirements and none have lead times that make it tough to respond. This analysis should show that all the work I did on EOQ and reorder points were right, and the real problem lies with those sales and finance people who don't understand our production needs."

CASE ANALYSIS

1. What are the key issues brought about in the conversation? What are the key symptoms and what are the underlying problems? Be specific in your answers.
2. Use the product information to develop an MRP approach to the problems. Would MRP solve the problems? If so, show specifically how MRP would avoid the problems as discussed by Ken and Jack.
3. Are there any conditions that would bother you about the ability of MRP to deal with the problems? What specifically are those conditions?
4. Suppose it was discovered that there were only 250 of the component E in stock instead of the 300 listed on the inventory record. What problems (if any), would this cause and what are some of the ways that these problems could be addressed? How would MRP help you (if at all) when other methods might not?
5. Suppose the design engineer advises that they have a new design for component F. It won't be ready until sometime after week 2, but they want you to give a date for the first supplier shipment to come in, and to be ready to tell the supplier how many to ship. Since the change is transparent to the customer, the design engineer's advice is to go ahead and use up any existing material of the old model. How will MRP help you to deal with this issue?
6. Can you think of any other "what if . . . " questions that might be more easily addressed by a systematic approach such as MRP?

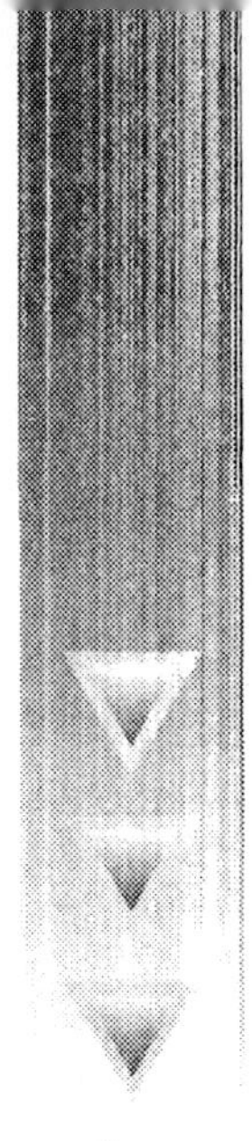

CASE 8

WESCOTT PRODUCTS

Whenever Jason Roberts thought about going to work on Friday morning, he started to get a little knot in his stomach. Jason had recently accepted the job as operations manager for a small manufacturing company that specialized in a line of assemble-to-order products. When he accepted the job he had been a recent graduate of a business program in which he specialized in operations. He had done fairly well in his classes and had emerged as a confident, self-assured person who was sure he could handle such a job in a small company.

The company, Wescott Products, had recently experienced rapid growth from the original start in a two-car garage just five years earlier. In fact, Jason was the first person they had ever named as operations manager. Prior to that, the only production "manager" reporting to the owner (Judy Wescott) was Frank Adams, the production supervisor. While Frank was an experienced supervisor, he had been promoted to supervisor directly from his old job as a machine operator and had no formal training in planning and control. He soon found that planning was too complex and difficult to handle for him, especially since he also had full responsibility for all the Wescott workers and their equipment. Randy Stockard, the sales and marketing manager, had requested and finally applauded Judy Wescott's decision to hire Jason, since he felt production was having a much more difficult time in promising and delivering customer orders. Randy was starting to spend more and more time on the phone with angry customers when they didn't get their order at the time they expected it. The time away from developing new sales and the danger of losing established customers started to make Randy highly concerned about sustaining sales growth, to say nothing about his potential bonus check tied to new sales!

Once Jason was placed in the position, however, the "honeymoon" was short and soon Jason started doubting how much he really did know. The company was still having trouble with promising customer orders and having the capacity to meet those orders. At first he thought it was the forecasting method he used, but a recent analysis told him the total actual orders were generally within 10% of what the forecast projected. In addition, production never seemed to have any significant shortages in either subassemblies or components. In fact, many felt they had far too much material, and in the last couple of staff meetings Jake Marris (the company controller) was grumbling that he thought the inventory turn ratio of just less than 3.5 was unreasonable and costing the company a lot of money. It must be something else, and he had to discover it quickly.

The first thing Jason thought about was to request the assembly areas to work overtime, but he soon found out that was a sensitive topic that was to be used as a last resort. The workers in that area were highly skilled and would be difficult, if not impossible to replace in any reasonable time. Adding more would also be difficult for the same reason. A year earlier they were being worked a lot of overtime, but had finally had enough. Even though Wescott had no union, the workers got together and demanded better overtime control or they would all quit to move to other jobs which were plentiful for skilled workers in this area. The agreement was that they were to be asked for no more than four hours per worker per week unless it was truly an emergency situation. They were well paid and all had families, and the time with their families was worth more to them than additional overtime pay. At least the high skill level had one advantage. Each of the workers in the assembly area could skillfully assemble any of the models, and the equipment each had was flexible enough to handle all the models.

On Friday mornings he made his master schedule for the next week (since the standard lead time for all assemblies was quoted as one week, Jason had felt no need to schedule further into the future when very few orders existed there), and no matter how hard he tried he never seemed to be able to get it right. He was sure that he had to start the process by loading the jobs that were missed in the current week into the Monday and Tuesday time blocks, and hope that production could catch up with those in addition to the new jobs that were already promised. The promises came when Randy would inform him of a customer request and ask for a promise date—which was often "as soon as possible." Jason would look at the order to see if they had the material to make it and if the equipment to make it was running. He would then typically promise to have it available when requested. Now that a lot of promises were not being met, however, Randy was starting to demand that Jason "get control" of the operation. Jason tried to respond by scheduling a lot of each model to be run every week, but he often found he had to break into the run of a lot to respond to expediting from sales. He knew this made matters worse by using extra time to set up the equipment, but what else could he do? Even Judy Wescott was asking him what she needed to do to help him improve the performance. His normal high level of self-confidence was being shaken.

Jason started pouring over his old operations book looking for something he could use. He finally realized that what he needed was a more effective system to develop master schedules from which he could promise orders, order components, and plan capacity. Unfortunately, he also recalled that when they covered that material in his class he had taken off early for spring break! Even though he knew enough to recognize the nature of the problem, he didn't know enough to set up such a schedule. Humbly, Jason called his old instructor to ask for advice. Once his instructor was briefed on the problem, she told him to gather some information that he could use to develop a sample master schedule and rough-cut capacity plan. Once he had the information, she would help show him how to use it.

The following describes what she asked him to collect:

1. Pick a work center or piece of equipment that has caused some capacity problem in the recent past. List all the product models that use that work center.
2. For each of the models, list the amount of run time they use the work center per item. Also list the setup time, if any. These times can be gathered from standards or, if the standard data are suspect in their accuracy or do not exist, use the actual average time from recent production.
3. For each of the models, list the usual lot size. This should be the same lot size used for the master schedule.
4. For each of the models, list the current inventory, the current forecast, and the current firm customer order quantities.
5. Compute the current capacity (hours) available for the equipment used.

The following tables summarize the data he collected:

Work Center 12

Model	Run Time (per item, in minutes)	Setup Time (per lot)	Lot Size (minimum qty.)	*On-hand
A	3.7	90 minutes	150	10
B	5.1	40 minutes	100	0
C	4.3	60 minutes	120	0
D	8.4	200 minutes	350	22
E	11.2	120 minutes	400	153

*Most of the on-hand was really forced when the lot size exceeded orders for the week for that model. The workers would then assemble the rest of the lot as "plain vanilla," such that they could easily add any subassembly options once the actual customer orders came in.

There are currently two workers assigned to the work center, assigned only to the first shift. Even though assembly workers are very flexible, Jason cannot take workers from another assembly area, since those work centers are also behind and therefore appear to be equally overloaded.

Following are the forecast and customer orders for each of the five models assembled in work center 12:

Model	Weeks	1	2	3	4	5	6	7	8	9	10
A	Forecast	45	45	45	45	45	45	45	45	45	45
	Cust. Orders	53	41	22	15	4	7	2	0	0	0
B	Forecast	35	35	35	35	35	35	35	35	35	35
	Cust. Orders	66	40	31	30	17	6	2	0	0	0
C	Forecast	50	50	50	50	50	50	50	50	50	50
	Cust. Orders	52	43	33	21	14	4	7	1	0	0
D	Forecast	180	180	180	180	180	180	180	180	180	180
	Cust. Orders	277	190	178	132	94	51	12	7	9	2
E	Forecast	200	200	200	200	200	200	200	200	200	200
	Cust. Orders	223	174	185	109	74	36	12	2	0	0

Once Jason had gathered all the data, he immediately called his instructor, only to find out that by an ironic twist of fate she would be gone for more than a week on spring break! This leaves you to help Jason by completing the Case Analysis that follows.

CASE ANALYSIS

1. Discuss the nature and probable sources of the problem.
2. Examine the rough-cut capacity situation using the data Jason gathered. Discuss the results and how they are linked to the problems identified in question 1.
3. Use the information and your knowledge of the situation to develop a complete plan for Jason to use in the future. Part of this plan should be to build and demonstrate the approach to master scheduling for the data given in the case.

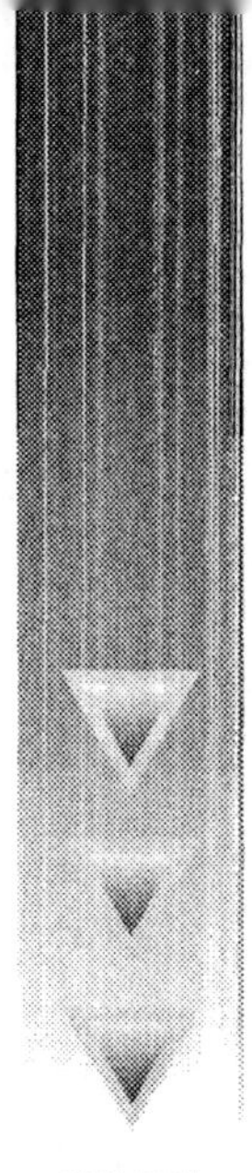

CASE 9

SANGSTER MANUFACTURING I: AGGREGATE PRODUCTION PLAN

Sangster Manufacturing makes a line of wheelbarrows consisting of three models: 4 cubic feet, 3 cubic feet, and 2 cubic feet. While the demand for each model varies, the same work centers are used in making each model. As well, some of the parts are common.

For planning purposes Sangster uses a four-week period. This provides for 13 periods of four weeks for the year. Based on past demand they have developed the following forecast for the next year:

Period	Demand Forecast	Period	Demand Forecast
1	500	8	1500
2	600	9	1000
3	900	10	800
4	1200	11	700
5	1400	12	600
6	1600	13	400
7	1800		

The cost of carrying inventory is calculated at $2 per unit per period. Because wheelbarrows are a common item and there is a lot of competition, the company feels that a stockout will result in lost sales. They estimate that the cost of a lost sale is approximately $20 a wheelbarrow. Consequently, it is company policy to carry a minimum inventory of 100 units at all times. At present they are carrying 500 wheelbarrows in inventory and they want to have the same number in stock at the end of the planning period.

Maximum production capacity is 1500 units per period in regular time and the plant is currently producing 800 units per period. Extra production, to a maximum of 100 units per period, can be added by working overtime. The overtime premium is $4 a unit more than regular time production. Some work can be subcontracted, but the subcontractor has limited facilities and can produce only 100 units per period. The cost of subcontracting adds $6 to the normal cost of a wheelbarrow. The rate of production can be changed at a cost of $6 a unit, which includes costs of hiring, training, and layoff.

CASE ANALYSIS

As the new materials manager you have been asked to develop a production plan for the next year that will meet the company's objectives and minimize the costs involved. Are there any other considerations aside from minimizing cost?

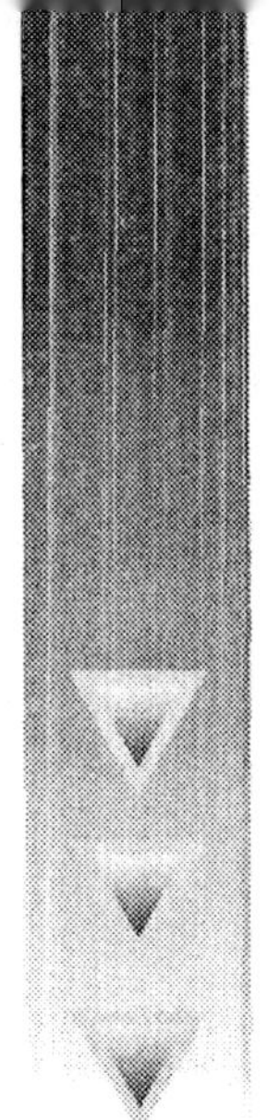

CASE 10

SANGSTER MANUFACTURING II: MASTER PRODUCTION SCHEDULE

Sangster Manufacturing makes a line of wheelbarrows consisting of three models: 4 cubic feet, 3 cubic feet, and 2 cubic feet. In Sangster Manufacturing Case I, you developed a production plan. Management has agreed to the following production plan for the next 13 periods:

Period	Demand	Production	Period	Demand	Production
1	500	100	8	1500	1500
2	600	600	9	1000	1000
3	900	900	10	800	800
4	1200	1500	11	700	800
5	1400	1500	12	600	800
6	1600	1500	13	400	500
7	1800	1500			

Historically, demand for the three models has been in the ratio of 20% (2 cubic foot), 45% (3 cubic foot), and 35% (4 cubic foot) of total demand. At present there are 125 two cubic foot, 175 three cubic foot, and 200 four cubic foot wheelbarrows in stock. Management insists on a minimum stock of 25 wheelbarrows in each size and a total minimum of 100 at any one time. There are to be no stockouts. The wheelbarrows are made in lots of 25. Setup cost is $200.

Case Analysis

Develop a master production schedule for the three models of wheelbarrows so that management's conditions are met and the setup cost is a minimum.

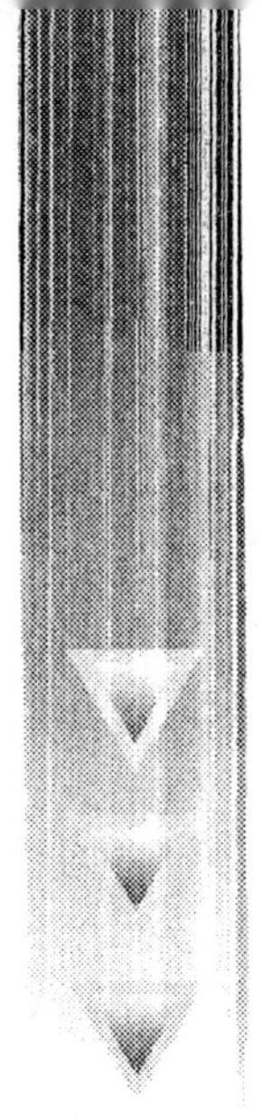

CASE 11

ATLANTIC ELECTRIC I: THE SURPRISE BACK ORDER

The Atlantic Electric Company supplies a wide range of electrical and electronic equipment to industrial equipment manufacturers and to the commercial market. The company was started in 1949 with the manufacture of a 35 amp disconnect switch suitable for residential wiring at that time. Since then, the number and sophistication of their commercial products have increased to include terminal blocks, multiphase disconnects, and circuit breaker panels. Commercial products are available off-the-shelf in a number of configurations which meet or exceed North American building codes. The industrial products have been developed from the ongoing requests by manufacturers to supply products that perform a specific function for a specific product. Their latest product was inspired by a major computer manufacturer. The 2001 Power Supply includes, along with the main electrical disconnect, the ability to regulate the voltage of the supply and to instantly switch to a backup power supply in the event of a power failure. Many of these new products such as the 2001, rely on electronic components.

Atlantic has experienced a growth in not only volume but also in the type of work done. They continue to bend and cut metal sheeting for the enclosers, form and rivet all the switch parts, and assemble and final test all the products. The latest products now require them to perform very detailed assembly of electronic subassemblies, which include some long lead time components. Purchasing has grown from local sourcing of metals and fasteners from a few suppliers to worldwide sourcing of electronic components. This has created more work for the purchasing department and has increased the lead time of their products.

In the past year, Atlantic has been experiencing problems with back orders of major products while at the same time work-in-process levels are at an all time high. Profits are not what they used to be and every department is under pressure to reduce costs and improve customer service.

The sales department has complained many times at recent meetings that production just doesn't seem to be able to produce what the customers want. At Wednesday's meeting Jack Adams, the national sales manager, complained that he had five major clients that were threatening to cancel their accounts if delivery didn't improve. At the meeting Jack had said, "One of my major clients had an overforecast overseas shipment to make and needed to ship as one unit. We could use this type of business and I was anxious to get the order. I told him that I would check every single item on the order and get back to him later the same day. The order had no custom products and included only off-the-shelf items, including thirty 42-Bs. Delivery was to be next week. We checked our master schedule for the major products including the 42-B before we took the orders and projected available inventory showed more than enough to cover these orders. The 42-B already had ten in stock and there were 100 to be put into stock this week. I even checked on the progress of the in-process order for 100 and was told that it was complete and about to be put into finished goods. We should see it on our screen as 'on hand' later that day. But, later that same day, when we entered the order for the 42-B we were told that there would be back orders. How can we work with a system that one minute tells us that there are products available and then tells us we can't ship these same products?" Janet Johnson, the materials manager, said she would personally look into the problem to see if there was product available for customers that the system wasn't recognizing and to see what could be done to still ship this order on time.

Later that day Janet met with her staff from Master Scheduling, Purchasing, and Production Control. She opened the meeting with the question, "What seems to be the problem here? Past production records don't show a capacity problem and even the forecast seems to be okay, for once. Why is this becoming a more frequent problem? WIP and overtime accounts are over budget and yet back orders are increasing."

The responses were quick to follow with claims that the sales department doesn't seem to follow its own forecasts and that substitutions were made without any thought about the lead time of the new products. Changes were being made too frequently and production always took the blame for not being able to produce the right materials at the right time. The computer system seemed to be working okay, but some people were abusing their access to information and often not understanding the consequences of their actions. Janet handed out copies of the current master schedule for the 42-B and gave everyone a few minutes to detect any problems.

Master Schedule June 16:42-B

Week		26	27	28	29	30	31	32	33	34	35	36	37
Forecast		20	25	20	30	25	25	30	20	25	30	20	30
Customer Orders		22	20	25	18	8		14		6		7	
Projected Available Balance	10	88	63	38	8	83	58	28	8	83	53	33	3
Available-to-Promise		25				78				87			
Master Production Schedule		100				100				100			

Wayne Donner, the master scheduler, was first to reply that he didn't see anything wrong with the schedule. Forecast and customer orders were not out of line right now and that total planned production was greater than booked orders. He did point out that production lead time for this product was only a matter of a few days, but the product relied on a specific transistor that had a four-week lead time and there was no way that an order for the extra products could be made within the next week. He pointed out that everything was in place for the next MPS of this product in a little over four weeks and asked if it would be possible for the sales department to wait until the next order would be ready.

Dale Wilson, the buyer, said that he could expedite the needed transistor, but it was impossible for him to get the needed component in less than two weeks. He was reluctant to change the order quantity of 100, which gave him a favorable discount. He also added that he had been working with this specific supplier to help keep costs down. "If we keep changing the order quantity and the delivery schedule this is only going to pass our problems onto the supplier. We have worked long and hard to mesh our operations with our suppliers. What can we do if we tell them that the information we are using keeps changing every day?" "Another thing. Say I could get the 30 high-amperage power transistors that we need? That's only going to mess up Steve's schedule and bump another order in production. You were complaining about work-in-process inventories."

"Hey, since when are you on my side?" said Steve Keenan, the production scheduler. "Dale's right though. It was okay back when we made the Atlantic 100 line. I could change from one order to another in no time at all. With our new and diversified production areas, we need some kind of warning before we make changes to the schedule. The best way to keep costs down in my department is to have some visibility. Then we can combine and anticipate setups. As for the 42-Bs I could have them out of here next week if I could get the parts. It will interfere with some other orders though, and I may need to put on some overtime. Let me know what you want to do. But, I will point out that I can't do a thing without the proper components. That transistor is not used in any other products and I couldn't steal from another order even if I wanted to."

Janet ended the meeting by telling everyone that she wanted their full cooperation on this order and to keep in touch with each other should any new developments occur. She asked Dale to contact the supplier of the necessary transistor and to let everyone else know the earliest anticipated delivery date. She also suggested that he fully review the bill of material for the 42-B and see if there were any other components that might delay shipment. She also mentioned that she planned to meet with customer order entry in Jack's department to see if they had any problems working with the system and if there was anything they could suggest to get this order out on time. She also asked if Wayne would like to attend the meeting to make sure that the entering of customer orders into the master schedule was being done according to approved procedures.

CASE ANALYSIS

1. Was Jack Adams misled by the information available in determining the ability to fill the order?
2. If all components for an order of 100 42-Bs could be made available in week 28 could they ship an additional 30 for Jack's overseas order? Recalculate the MPS with this information to support your answer.
3. What guidelines would you recommend Atlantic Electric implement to help manage changes to the MPS?
4. Discuss the use of safety stock for the 42-Bs. What amount would you recommend and how would this amount change the initial situation?

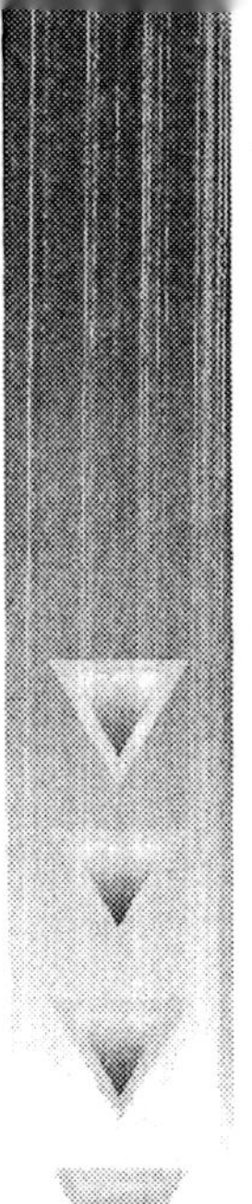

CASE 12

ATLANTIC ELECTRIC II: THE SHORT SHIPMENT

Atlantic Electric has a metal-forming plant in San Antonio, Texas. Atlantic makes a wide variety of commercial and industrial electrical products, used to disconnect and protect wiring systems including computer systems. They have an MRP II system in place and use the system to release work orders to manufacturing and recommended quantities to purchasing.

Two of the products manufactured at the San Antonio location are standard metal enclosures for electrical disconnects that are made from metal sheet, which comes in 2,000-pound rolls. Previous work in standardizing their products has lead to the ability to make both final products (A and B) entirely from one raw material (X). There are some components (W, Z and Y) that go into the final products. These are manufactured in separate areas and are made from raw material (X). All information on the final products, components, and raw material are given in typical MRP format. All the information is stated in the attached item master files, where-used files, and master production schedules. You are the planner in charge of producing the As and Bs plus all the components and the raw material. (Note that part Y experiences some independent demand. It is a panel used in joining one or more enclosures together by the installer and there is a small continuous demand for this part.)

ASSIGNMENT

Complete the planning sheets for all the parts according to the steps below.

1. Complete the master schedules for parts A and B.
2. Using the information from the where-used files, construct product trees for the final products.
3. For each product enter the low-level code on the respective item master file.
4. Complete the MRP worksheets for each of the components and for the raw material. Be sure to include all of the information on each part including lead time offset from the master schedule, scrap factors, and lot sizes. A separate pegging file has been included for part X to accumulate the gross requirements.
5. What exception messages (if any) would be generated by the MRP system with the current situation?
6. You have just received a fax from the steel mill in Pittsburgh that supplies the high quality steel used in your product. An error was made in your last shipment and the quantity shipped via ocean freight was only 2,000 pounds. They will correct the error on your next order, however, you do not have any current orders with the mill. They will not be able to reduce the lead time of your next order, since they are very reluctant to change their schedule for such a small quantity of steel. Assume that you cannot get any material prior to week 8. Suggest what you might do to prevent any shortages of your final products according to the current schedule information.

CASE ANALYSIS

1. Discuss the costs and benefits of keeping safety stock of part Y compared to keeping safety stock of part X.
2. One option of dealing with the short shipment would be expediting the shipment via road transport rather than ocean freight. Discuss the costs and benefits of shipping via road for all future shipments.

Item Master File—Part A

Static Information	
Lead Time	2 weeks
Low-Level Code	
Lot Size	100 units
Safety Stock	0
Scrap factor%	
Dynamic Information	
On-Hand	100 units
On Order	0
Independent Demand	*MPS*

Item Master File—Part B

Static Information	
Lead Time	1 week
Low-Level Code	
Lot Size	200 units
Safety Stock	0
Scrap factor%	0
Dynamic Information	
On-Hand	100 units
On Order	0
Independent Demand	*MPS*

Item Master File—Part W

Static Information	
Lead Time	1 week
Low-Level Code	
Lot Size	L4L
Safety Stock	0
Scrap factor%	0

Dynamic Information	
On-Hand	0
On Order	100, week 1
Independent Demand	

Item Master File—Part X

Static Information	
Lead Time	8 weeks
Low-Level Code	
Lot Size	4,000 lbs
Safety Stock	0
Scrap factor %	0

Dynamic Information	
On-Hand	3,400 lbs
On Order	4,000, week 3
Independent Demand	

Item Master File—Part Y

Static Information	
Lead Time	2 weeks
Low-Level Code	
Lot Size	500 units
Safety Stock	100 units
Scrap factor %	0

Dynamic Information	
On-Hand	200 units
On Order	
Independent Demand	

Item Master File—Part Z

Static Information	
Lead Time	2 weeks
Low-Level Code	
Lot Size	L4L
Safety Stock	0
Scrap factor %	10

Dynamic Information	
On-Hand	200 units
On Order	200, week 2
Independent Demand	

Where-Used File—Part W

Parent	A
Quantity Per	1

Where-Used File—Part Z

Parent	A
Quantity Per	2

Where-Used File—Part Y

Parent	B
Quantity Per	3

Where-Used File—Part X

Parent	Z	B	W	Y
Quantity Per	1	3	3	2

Master Schedule

PART #

Week	1	2	3	4	5	6	7	8	9	10	11	12
Forecast	100		100		120		140		120		100	
Projected Available Balance												
MPS Receipts												

Master Schedule

PART #

Week	1	2	3	4	5	6	7	8	9	10	11	12
Forecast	80			80			100			120		
Projected Available Balance												
MPS Receipts												

MRP Record **Part #**

Week	1	2	3	4	5	6	7	8	9	10	11	12
Gross Requirements												
Scheduled Receipts												
Projected Available Balance												
Net Requirements												
Planned Order Receipts												
Planned Order Releases												

MRP Record **Part #**

Week		1	2	3	4	5	6	7	8	9	10	11	12
Gross Requirements													
Scheduled Receipts													
Projected Available Balance													
Net Requirements													
Planned Order Receipts													
Planned Order Releases													

MRP Record **Part #**

Week		1	2	3	4	5	6	7	8	9	10	11	12
Gross Requirements													
Scheduled Receipts													
Projected Available Balance													
Net Requirements													
Planned Order Receipts													
Planned Order Releases													

MRP Record **Part #**

Week		1	2	3	4	5	6	7	8	9	10	11	12
Gross Requirements													
Scheduled Receipts													
Projected Available Balance													
Net Requirements													
Planned Order Receipts													
Planned Order Releases													

Single-Level Pegging Record for PartX

Use the following table to accumulate the requirements for PartX.

Week	1	2	3	4	5	6	7	8	9	10	11	12
Parent												
Gross												
Requirements												

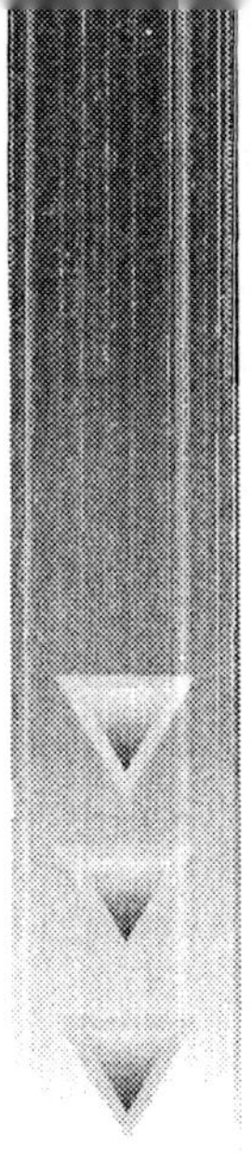

CASE 13

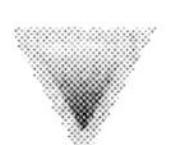

CONNERY MANUFACTURING

When Juan Hernandez was first given the position of head buyer for the Connery Manufacturing Company, he visualized the job as merely an expansion of his old position as a commodity buyer. He had no formal training when he took the position, having been promoted to commodity buyer from his position as inventory clerk. That job he had gotten directly from high school.

The lack of formal training was not a problem when he first took the job. The Connery Company was small but growing, and the major concern of the purchasing department was to obtain adequate purchased material to support the production and the growth in sales. There was little done in the way of price negotiation. The reasons were that there was little competition for their products and all costs could easily be passed along in the product price, leaving room for the healthy profits that have helped Connery grow so rapidly.

As is often the case in these types of situations, the luxury of little competition and flexibility in pricing was fairly short-lived. The success of the products Connery produced attracted a lot of attention, and soon Connery found itself in a market with several strong competitors.

Although they still had the advantage of some recognition in the market ("first mover" advantage), that advantageous position was in grave danger of erosion. They also had an advantage in being farther down the learning curve, and the quality of their product had always been quite good. The problem now was cost. Competition was driving down the prices and maintaining their delivery record at a lower price was rapidly becoming an important factor in stemming the tide of market share erosion.

These factors were one of the key reasons that Juan was promoted. He was recognized as the best and most experienced of all the buyers, and Mr. Connery recognized the need to move the procurement activity from one of passive buying into an active and aggressive supply management group.

As a buyer, Juan's primary responsibility was to get the material they needed, when they needed it. He primarily was responsible for buying standard components and materials, so he had hundreds of catalogs from all possible suppliers of these standard commodities. When he needed to place an order he would typically use the catalog price or the quoted price from a supplier as long as they could meet the delivery time he needed. He had little concern for transportation cost or even quality, since for these standard components the quality from all possible suppliers was roughly equivalent. There were a few cases in the past when quality did prove to be a problem, but the supplier could usually respond quickly with an appropriate replacement. Even though the supplier would typically give Connery credit for any rejected parts, changing schedules around the problem or carrying safety stock to protect against problems would both end up costing Connery more money.

Soon after the promotion to head buyer, Juan realized the job would be much more than merely an expansion of his old position. Mr. Connery told Juan he created the position of head buyer to move the company into a more cost-competitive condition. He wanted Juan to develop and implement a plan that would attempt to accomplish the following:

- Reduce raw material (purchased) inventory levels
- Improve delivery speed and reliability of purchased material
- Improve the quality performance of suppliers
- Reduce the overall cost of purchased materials

These actions were considered to be important if they were to reduce the overall cost and stay "ahead of the pack" on price competitiveness.

Juan now realized both the extent and the seriousness of the new position and his responsibility. The following give a little indication of the current position of the company:

Annual Cost of Goods Sold	$14,827,527
Direct Material Cost	$8,517,323
Inventory (on Balance Sheet)	$2,352,117
Supplied Parts Transportation Expense	$256,103
Number of Suppliers	2,872
Annual Inventory Holding Cost	21%
Average Total Processing Time for Products	3 hours, 27 minutes
Number of Different Designs for End Product	72

Case Analysis

1. What additional information should Juan gather to help him develop his plan? Explain how you would use the information.
2. Assuming you know the information, develop a plan for Juan.

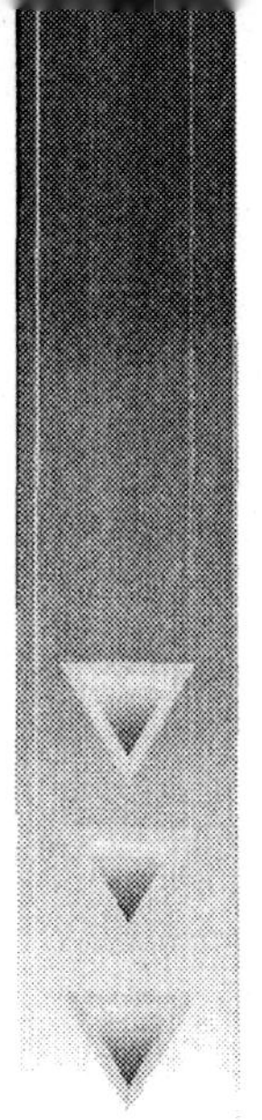
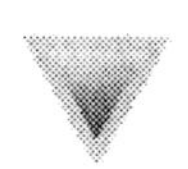

CASE 14

CONSOLIDATED WEEDCUTTERS

Consolidated Weedcutters Inc. makes two models of cutters, the standard (part number 100), and the economy (part number 150). The following information is taken from the bill of material files for each of the weedcutters.

Part Number	Made from Part Number	Quantity per
100	301	2
	302	2
150	302	1
	303	1
301	303	2
	304	1
302		
303	302	1
	304	1
304		

Part 100 is planned for using a master production schedule. Lead time is two weeks and it is built in lots of 200. At present there are 200 scheduled to be received in week 1. None is on hand.

Part 150 is planned for using a master production schedule. Lead time is one week and it is built in lots of 100. At present there are 100 scheduled to be received in week 1. None is on hand.

Part 301 is a component having a lead time of two weeks, and a lot size of 600. At present there are lots scheduled to be received in week 1 and week 2.

Part 302 is a component with a lead time of one week, and a lot size of 1500. At present there are 500 on hand and a lot to be received in week 1.

Part 303 is a component having a lead time of two weeks, and a lot size of 1500. At present there is a lot scheduled to be received in week 1.

Part 304 is a component having a lead time of four weeks, and a lot size of 2000. At present there are 2200 on hand and there is a lot scheduled to be received in week 4.

Case Analysis

1. From the information given, draw product trees for each of the finished goods.
2. What is the low-level code for each of the items?
3. Following are sheets for each part. Each sheet has an item master file, a where-used file, a pegged requirement file, and an MPS or MRP work sheet. These files are normally located in different spaces in a computer system, but are combined here for your convenience. Complete the work sheets, filling in information from what is given above, and performing the required calculations.
4. It is the beginning of week 1. What exception message(s) will the computer generate? What action(s), if any, should be taken?
5. The scheduled receipt for part 302 in week 1 has been received and is 100 parts short. What parts will this affect? What action should be taken?
6. An order is received for 25 of item 100 for delivery in week 3. Can these be promised without changing the MPS?

Part Number: 100

Item Master File

Planning Factors

Low-Level Code: Lead Time: Order Quantity: Safety Stock:

Status

On-Hand: On Order: Due Date:

Master Schedule

Week	1	2	3	4	5	6	7	8	9	10
Forecast Requirements	100	100	100	100	100	100	300	100	200	100
Customer Orders	95	90	90	110	70	50	190	60	120	30
Projected Available Balance										
Available-to-Promise										
Master Production Schedule										

Part Number: 150

Item Master File

Planning Factors

Low-Level Code: Lead Time: Order Quantity: Safety Stock:

Status

On-Hand: On Order: Due Date:

Master Schedule

Week	1	2	3	4	5	6	7	8	9	10
Forecast Requirements	50	50	50	50	150	50	50		50	50
Customer Orders	45	48	48	50	120	30	35	10	25	30
Projected Available Balance										
Available-to-Promise										
Master Production Schedule										

Part Number:

Item Master File

Planning Factors

Low-Level Code: Lead Time: Order Quantity: Safety Stock:

Status

On-Hand: On Order: Due Date:

Where-Used File

Where Used:

Quantity Per:

Pegging File

Week		1	2	3	4	5	6	7	8	9	10
Gross Requirements	Part										

Material Requirements Plan

Week		1	2	3	4	5	6	7	8	9	10
Gross Requirements											
Scheduled Receipts											
Projected Available Balance											
Net Requirements											
Planned Order Receipts											
Planned Order Releases											

Part Number:

Item Master File

Planning Factors

Low-Level Code: Lead Time: Order Quantity: Safety Stock:

Status

On-Hand: On Order: Due Date:

Where-Used File

Where Used

Quantity Per

Pegging File

Week		1	2	3	4	5	6	7	8	9	10
Gross Requirements	Part										

Material Requirements Plan

Week		1	2	3	4	5	6	7	8	9	10
Gross Requirements											
Scheduled Receipts											
Projected Available Balance											
Net Requirements											
Planned Order Receipts											
Planned Order Releases											

Part Number:

Item Master File

Planning Factors

Low-Level Code: Lead Time: Order Quantity: Safety Stock:

Status

On-Hand: On Order: Due Date:

Where-Used File

Where Used:

Quantity Per:

Pegging File

Week		1	2	3	4	5	6	7	8	9	10
Gross Requirements	Part										

Material Requirements Plan

Week		1	2	3	4	5	6	7	8	9	10
Gross Requirements											
Scheduled Receipts											
Projected Available Balance											
Net Requirements											
Planned Order Receipts											
Planned Order Releases											

Part Number:

Item Master File

Planning Factors

Low-Level Code: Lead Time: Order Quantity: Safety Stock:

Status

On-Hand: On Order: Due Date:

Where-Used File

Where Used:

Quantity Per:

Pegging File

Week		1	2	3	4	5	6	7	8	9	10
Gross Requirements	Part										

Material Requirements Plan

Week		1	2	3	4	5	6	7	8	9	10
Gross Requirements											
Scheduled Receipts											
Projected Available Balance											
Net Requirements											
Planned Order Receipts											
Planned Order Releases											

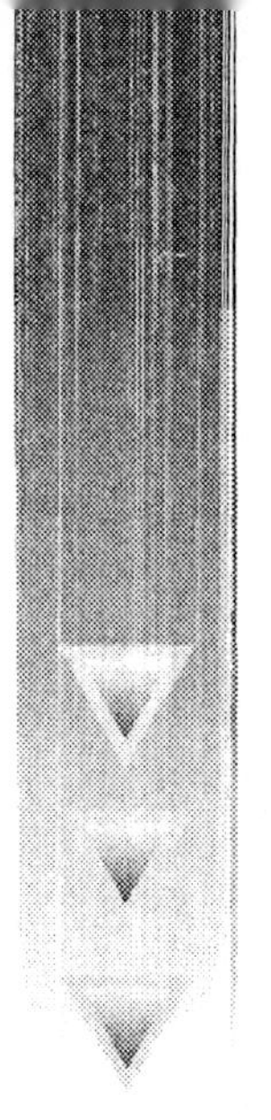

CASE 15

THE CASE OF THE VANISHING INVENTORY

Nick Talbot had a problem. As the master planner for Hardy Custom Engraving he was responsible for ensuring that they meet the sometimes conflicting priorities of sales, production, and finance. Something has gone terribly wrong, and all Nick's plans have failed. Despite his best efforts to level production and still meet promised delivery dates, two customers will not be getting their shipments this week. Jim "Buster" Hush had just pointed out the obvious by declaring in front of everyone in the meeting that, "You should have seen it coming!!", but Nick couldn't see how.

Nick stared at the latest output from the MRP system on his desk for K-1010, the product that was ruining his career. How could things have gone so wrong in two weeks? When Leona Hardy had taken over the company from her father, she came with some new ideas, many of which were already paying big dividends, and the company was really starting to grow. One of these ideas was the introduction of wood products to the existing line of metal fabricated items. There is no arguing that the line is proving popular with customers (and the salesforce) resulting in higher revenues and profits. Demand for K-1010 is a perfect example of this. Who would have thought that there was such a big market for laser engraved hardwood key chains? Well, the report told the tale.

Hardy had been selling the new key chains for the last 15 weeks or so. Demand started slow, but grew steadily. Everything had been going according to plan. There had been no problems with the supplier of the oak stock from which the key chains were cut. The line had reported no problems. No shortages, no production problems, but no finished goods when the system was telling him there should be ample coverage for the current customer orders. They just weren't there.

When he first got wind of the problem, Nick had thought that sales might be up to their old tricks of booking new business inside the frozen zone. He had floated that in the meeting without giving it much thought. That had really lit Buster's fuse. Buster was adamant that no new orders were being booked inside the lead time for the key chains. He had already heard from Leona that those kinds of actions were not going to be tolerated. He remembered her distinctly saying, "This is a new product line and we wanted to make sure that we don't start off by making promises we can't keep." Buster knew it wouldn't be wise to test her on this (at least not right away), so he had instructed all his reps to pay very close attention to the available-to-promise values. He knew if there were any arguments down the line, he could claim he depended on the information system.

In looking at the MRP record, Nick wondered why there were no material shortages for the oak stock. He knew that there were occasional knots and off cuts that meant that there should have been some scrap or yield issues but there was no scrap factor for the key chains. How was it that no material usage variances were being reported. He reviewed the issues of the oak stock from stores over the last couple of months hoping that would give him some clue. While future products would use oak, the key chains are currently the only product to use it, at a rate of 15 linear feet per 100 key chains. The latest cycle count, taken this morning, showed 1650 linear feet of oak on the shelf.

CASE ANALYSIS

1. Based on the information given, what is causing the poor customer service?
2. What should Nick do to correct the problem for the immediate future?
3. What changes would you make to the system information to avoid problems in the long run?

MRP Detail Output

Printed Production Week #21

PART NUMBER: K-1010 Keychain - Oak

LEAD TIME (WKS):	2.00	Lot Size:	1000
ON HAND:	300	Scrap Rate:	0
SAFETY STOCK:	0		
LATE RELEASES:	0		
# RECOM EXPEDITE:	0		
#RECOM DE-EXPED:	3800		

Production WEEK #	Past Due	21	22	23	24	25	26
GROSS REQUIREMENTS	0	0	10000	16500	0	14300	0
SCHEDULED RECEIPTS	3800	0	11000	0	0	0	0
PROJECTED AVAILABLE	4100	4100	5100	600	600	300	300
NET REQUIREMENTS	0	0	0	11400	0	13700	0
PLANNED ORDER RECEIPTS	0	0	0	12000	0	14000	0
PLANNED ORDER RELEASE	0	12000	0	14000	0	0	0

MRP Detail Output

Printed Production Week #23

PART NUMBER: K-1010 Keychain - Oak

LEAD TIME (WKS):	2.00	Lot Size:	1000
ON HAND:	900	Scrap Rate:	0
SAFETY STOCK:	0		
LATE RELEASES:	0		
# RECOM EXPEDITE:	0		
#RECOM DE-EXPED:	4200		

Production WEEK #	Past Due	23	24	25	26	27	28
GROSS REQUIREMENTS	0	16500	0	14300	0	15400	12200
SCHEDULED RECEIPTS	4200	12000	0	0	0	0	0
PROJECTED AVAILABLE	5100	600	600	300	300	900	700
NET REQUIREMENTS	0	0	0	13700	0	15100	11300
PLANNED ORDER RECEIPTS	0	0	0	14000	0	16000	12000
PLANNED ORDER RELEASE	0	14000	0	16000	12000	0	0

Production Orders Report
Printed Production Week #23

PART NUMBER: K-1010 Keychain - Oak

Order #	Status (PL/OP/CL)	Release Week #	Required Week	Release Quantity	Completed to Date
03071	OP	07	09	4000	3450
03264	OP	11	13	6000	5300
03281	OP	12	14	9000	8150
03503	OP	14	16	9000	7975
03977	OP	18	20	10000	9325
04106	OP	20	22	11000	10600
04332	OP	21	23	12000	0
	PL	23	25	14000	0
	PL	25	27	16000	0
	PL	26	28	12000	0

Storeroom Transaction Report
Material Receipts and Issues
Printed Week #23

Part# OA-150 - Oak Stock 1" x 0.25" Finished Grade

TRANX#	Week	ISS/REC	Order #	Qty	UM
213024	07	R	P-08733	3600	ft
213135	07	I	W-03071	600	ft
218070	11	I	W-03264	900	ft
221010	12	I	W-03281	1350	ft
222907	14	R	P-09076	3600	ft
223006	14	I	W-03503	1350	ft
224813	18	I	W-03977	1500	ft
226577	20	R	P-09213	3600	ft
226641	20	I	W-04106	1650	ft
226792	21	I	W-04332	1800	ft

TOTAL NUMBER OF UNITS ISSUED: 9150
TOTAL NUMBER OF UNITS RECEIVED: 10800

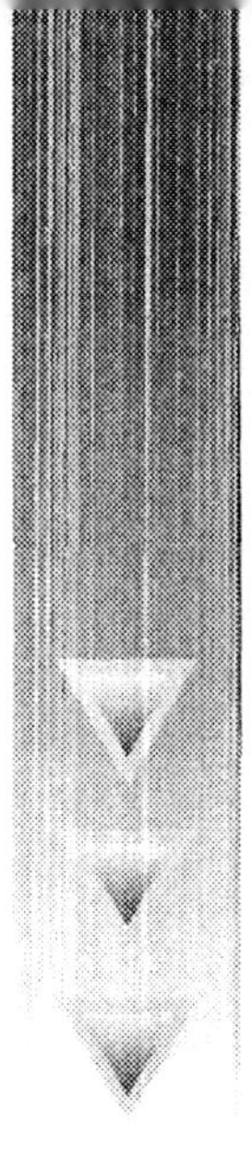

CASE 16

YOUR OFFICE WAREHOUSE

John Caprelli was getting a little concerned about the inventory situation in his successful and growing office supply business called "Your Office Warehouse." While he did not always have the cheapest prices in the area, he grew his business on having virtually anything any office might need, and he would guarantee delivery to any business within 50 miles in three hours or less. Needless to say, the delivery costs for John were high, but there was little he could do about that since that aspect was a significant competitive advantage for him.

What was bothering John was the increasing cost of inventory. He felt he had to keep plenty of inventory on hand, since he had to maintain his competitive advantage, but the inventory was costing him a large amount of money. First, some of the inventory was subject to becoming spoiled or obsolete. Just last month he had to scrap a case of printer's ink that had reached its expiration date, he had to give away two cases of paper that had become damaged from moving it too often, and had discovered some of the computer software had become obsolete when a new revision had been issued before he sold the older version. Secondly, he had to keep very close track of the inventory to ensure that he did not run out before he even knew there was a problem. The distributor that supplied him was good and quite reliable, but generally wanted one week lead time to deliver. If John did have to expedite some material, it would generally cost him a very large delivery premium. Even with the point-of-sale computer terminals John used for "cash registers," inventory still had to be constantly checked. Mislabeling, accidentally placing material in the wrong location, and some inevitable shoplifting all took an agonizing toll on the inventory accuracy. In the past, some of his people suggested he could help the situation by handling some inventory differently from others, but since they could not give him specifics as to how to do this, he continued to treat all inventory the same.

One of John's part-time cashiers (Ruth) was a full-time operations management student in the local university. She had told John that he should consider using ABC inventory principles together with more effective safety stock principles to reduce his cost while still maintaining his reputation for customer service and delivery. Once she had explained a little more, John became interested and relieved her of her cashier's duty for the next month so that she could analyze and possibly set up the inventory in the way she described to John.

Before Ruth had a complete "green light" to change the entire system, John asked her to take a subset of the inventory and show him quantitatively how it might save money and still be better for their business focus on availability. Table 1 summarizes her findings for the subset of the inventory.

John's initial policy was to try to reorder if the inventory reached one week's worth, but a rash of stockouts forced him to quickly reconsider. The current policy is to reorder with about two weeks left to account for the accuracy problems. With the current interest rates and other costs John had, he generally thought the inventory holding cost was about 20% of the value of the product per year. In addition, he had a policy of trying to maintain a 95% customer service level.

CASE ANALYSIS

1. Develop a reorder policy using the information above that would represent the "ideal" situation (assuming perfect information accuracy). Compare the cost of that policy with the annual cost of ordering the 30 items listed in Table 1 using John's current policy.
2. How would you use ABC information to improve the situation with respect to handling the inventory and the inventory accuracy problems? Establish a specific set of policies John and Ruth should use. Why is the method you suggest preferable to the current situation?

Table 1 Subset Inventory Findings

Part Number	Average Weekly Sales	Weekly Standard Deviation of Demand	Item Value in Dollars
A127	15	3	$12.00
A144	124	23	$1.50
A247	330	41	$0.24
B188	91	12	$3.76
B381	35	14	$5.22
B475	8	1	$61.00
B613	107	11	$73.08
B875	3	1	$164.55
B923	56	15	$31.90
C142	241	16	$14.88
C185	93	10	$6.53
C216	72	9	$18.24
C301	554	67	$0.33
C566	145	16	$2.44
C602	178	25	$5.43
C664	21	5	$16.78
C791	101	36	$9.54
C973	216	27	$11.03
D291	13	2	$87.90
D452	88	7	$117.23
D523	31	15	$0.17
D747	12	2	$23.44
D990	125	21	$53.87
E111	65	17	$78.21
E327	4	1	$258.70
E452	85	30	$46.66
E612	231	16	$25.40
E775	22	4	$48.20
E790	101	17	$1.34
E906	76	19	$12.55

CASE 17

SANDBAR BEACH HOTEL

The Sandbar Beach Hotel is a family business catering to both a holiday and a business market. It has 100 rooms, facilities for small conferences, and good recreational facilities.

June, July, and August are peak months during which the hotel has to turn away potential guests. Average room rates are $100 per night throughout the year, but vary somewhat with the seasons. Direct operating costs have averaged 60% of sales over the past few years.

It is January 1999. Because of the increase in business, management is considering whether they should expand. Management has gathered the data shown in Tables 1 and 2. Table 1 shows that room occupancy has increased steadily over the past seven years. It is expected that business will expand at the same rate for the next period of years. Table 2 shows the room occupancy by month for the past three years.

Table 1 Yearly room occupancy

Year	Rooms Occupied
1992	11,000
1993	11,800
1994	13,300
1995	14,500
1996	16,699
1997	18,037
1998	19,558

Table 2 Room occupancy per night			
Month	**1996**	**1997**	**1998**
Jan	33	35	38
Feb	12	15	18
Mar	25	30	33
Apr	45	50	52
May	60	70	75
June	75	80	90
July	95	99	100
Aug	95	99	100
Sept	45	50	55
Oct	12	13	25
Nov	12	12	15
Dec	40	40	42

Management is considering four options: stay as is; add 30 rooms; add 35 rooms; or add 40 rooms. The cost of expansion has been calculated to be $200,000 plus $25,000 per room added. They have asked you to recommend a course of action.

CASE ANALYSIS

Forecast the demand for rooms for the next five years. On that basis, calculate the expected increase in profit over this period for each of the alternatives. What is the payback period for each option? Make your recommendation. What other factors might be considered?

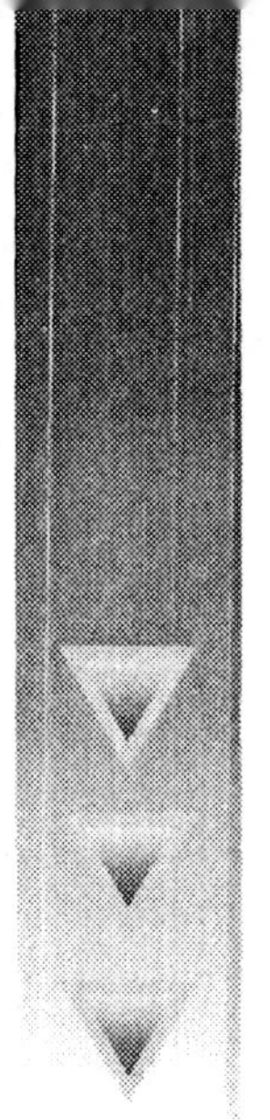

CASE 18

CROFTS PRINTING COMPANY

John Burton was not a happy man. He was a supervisor for the Crofts Printing Company, having been recently promoted from lead printer. While he felt very comfortable with his knowledge and success in the printing business, this managerial position was starting to wear on him. John was determined not to let it get him down, however, as he felt he surely had the knowledge, experience, and respect of the workers. He had been asking Jason Crofts for months for a chance at management, and he certainly wasn't about to let the job get the better of him.

THE CURRENT SITUATION

Since John had become supervisor the salespeople always asked him about an order before they promised delivery to a customer. He thought that would be quite simple—after all, who knew more about the printing business than he did? Based on his knowledge of the processes and what was already in progress, he gave what he thought were reasonable, even conservative, estimates of promise dates. Unfortunately, his "track record" was not too good. There had been many late deliveries since his managerial appointment, and nobody in the organization was too happy about it.

At first he thought the problem lay with the other workers. "They're just jealous about my selection as supervisor and want to make me look bad," was his initial reaction. Henry Hurley, another longtime machine operator, was John's best friend. One afternoon over a beer John asked Henry about the problem in a confidential discussion. Henry said he was sure that John had been trying to get it right but somehow it didn't seem to be going well. Henry assured John that the workers were trying their best. In fact, according to Henry, the workers had been putting in extra effort. They viewed John's promotion as a positive sign that there was a possible future for them in management as well. John's failure would have been, in fact, greatly discouraging to most of the workers.

John then thought he might be the problem when it came to giving estimates. The salespeople would almost always contact him about a possible job, to ask him when it should be promised to the customer. His great knowledge of the printing business allowed him, he thought, to quickly come to a good estimate. Perhaps he was not as good at estimating as he thought. To check this out, he looked at most of the jobs done during the last couple of weeks. In almost every case, the work recorded against a job was almost exactly what he had estimated. What little error existed was certainly not large enough to cause the problem.

John trusted Henry and believed his account of the situation, and his analysis of the estimates convinced him the problem wasn't there. If it wasn't the workers and wasn't the estimates, then what could it be? He must do something. Jason Crofts was a patient man, but there was a limit. He was worried about alienating his best customers, and at the same time knew he must be concerned about efficiency as a way to control cost.

John decided there was a need to take drastic action to ease the situation, or at least to find out what the cause was. On a Friday, he scheduled overtime for Saturday to finish all jobs in progress. On Monday, therefore, he could start with a clean slate. There were several jobs already promised, but not yet started. He figured that on Monday he could start with all new jobs and really figure out the source of his problem.

The jobs promised were all within four days, but he figured there should be no problem. He had three operations, and most of the jobs went through all three, but not all jobs needed all operations. He had one worker assigned to each operation. Over the next four days that represented 96 hours of available work time (3 operations times 4 days times 8 hours per day), and he had eight jobs promised. The total estimated time for all eight jobs was only 88 hours, giving him a buffer of 8 hours over the next four days. To make sure there would be no problem, he decided there would be no new jobs scheduled to start during those four days, with the only exception being if any worker completed all the necessary operations for all eight jobs before the end of the four days, they could start another. In any case, he wanted to make sure that if necessary all 96 hours would be reserved for just the 88 hours of scheduled work.

John had learned that a good priority rule to use was the critical ratio rule, primarily because it took into account both the customer due date and the amount of processing time for a job. He therefore used that rule to order job priorities. The following table shows the eight jobs, together with processing time estimates and due dates. All due dates are at the end of the day indicated. Processing times for all jobs at all three operations are in hours.

Job	Operation 1	Operation 2	Operation 3	Total Time	Day Due
A	5 hours	3 hours	4 hours	12 hours	Tuesday
B	0	6	2	8	Wednesday
C	4	2	5	11	Tuesday
D	7	0	3	10	Thursday
E	2	8	0	10	Thursday
F	0	6	3	9	Thursday
G	3	3	5	11	Thursday
H	6	5	6	17	Thursday

CASE ANALYSIS

1. Using the critical ratio rule, establish the priority for the eight jobs.
2. Use a Gantt chart to load the operation according to the priority rule established. In other words, load the most important job in all three work centers, then the next most important, and so forth. This is the method that John used.
3. Analyze John's approach and try to determine if he has a problem, and if he does determine the source of the problem.
4. Try to provide a solution for John that will ease the problem, and perhaps eliminate it.

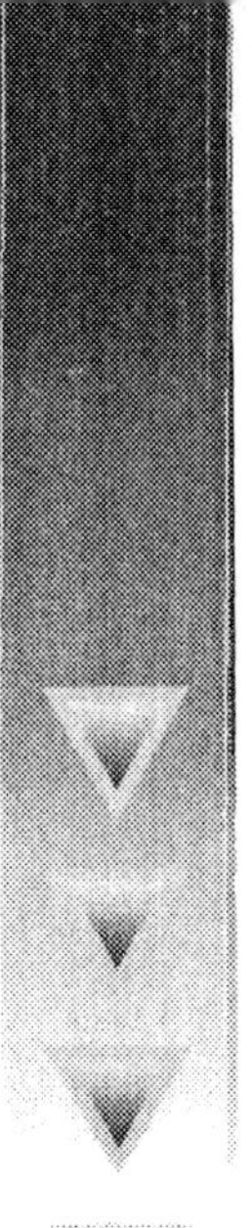

CASE 19

CARL'S COMPUTERS

There was no question about Carl's genius. Seven years ago he decided to enter the competitive nightmare that the personal computer business had become. While on the surface that appeared to be a rather non-genius-like move, the genius came in the unique designs and features that he developed for his computer. He also figured a way to promise delivery in only two days for the local and regional market. Other computer makers also had rapid production and delivery, but they were national competitors and the delivery time from distant locations generally made Carl able to out-compete them on delivery.

Carl soon had a loyal following, especially among the many small businesses in the area. Not only could Carl deliver quickly, but he also had very rapid service to deal with any technical problems. That service feature became critical for the local businesses whose very livelihood depended on the computers, and soon that rapid service capability became more important than the initial product delivery. Since most of these businesses were fairly small, they could not afford to have their own in-house computer experts, therefore they depended heavily on Carl.

THE CURRENT SITUATION

All was not totally rosy at Carl's Computers, however. Recently they had hired Rosa Change for the newly developed position of Inventory Manager for Aftermarket Service. In the first week Rosa got a good idea of the challenges facing her after she interviewed several of the people at Carl's.

Randy Smith, Customer Service Manager, "I'm not sure what you need to do, but whatever it is needs to be done fast! At this point our main competitive edge other than product delivery is service response, and I'm always hearing that we can't get a unit in the field serviced because some critical part is missing. Both the customers and the field service people are complaining about it. They make a service call, find out they need a certain part, but in many cases we're out of that part. The customers tend to be fairly loyal, but their patience is wearing thin—our policy is to provide at least a 98% customer service level, and we're not even close. That's not the only problem, though. Since our service is declining, the customers are looking more closely at our prices. I'd like to cut them a break, but our financial people tell us our margins are already too thin and get this—one major reason is that our inventory and associated inventory costs are too high! It looks to me as if we have a very large amount of the wrong stuff here. I don't know that for sure, but I sure hope you can find a solution, and fast!"

Ellen Bedrosian, Chief Engineer, "Boy, am I glad you're here! The inventory problems are killing us in engineering. Carl's has always been known for unique designs, and we've been trying hard to keep ahead of the competitive curve on that issue. The problem is that most of the time when we push hard to get a new design out, the inventory and financial people tell us we have to wait. It seems like they always have too much of the old design inventory around, and the financial "hit" to make it immediately obsolete would be too severe. We're told that as soon as we announce a new design many of our customers would want it, so that tends to make most existing old design material—even for service—obsolete. We try to tell the service inventory people when we have a new design coming so they can use up the old material, but somehow it never seems to work out."

Jim Hughes, Purchasing Manager, "Well, Rosa, I wish you luck—you'll need it. I'm getting pressure from so many directions sometimes I don't know how to respond. First, the financial people are always telling me to cut or control costs. The engineers then are always coming out with new designs, most of which represent purchased parts. A lot of our time is spent working with suppliers on the new designs, while trying to get them to have very rapid delivery with low prices. While most can live with that, where we really jerk them around is with the changes in orders. One minute our field service people tell us they've run out of something and they need delivery immediately. In many cases they don't even have an order for that part on the books. The next thing you know they want us to cancel an order for something that only a day before they said was critical. Our buyers and suppliers are good, but they're not miracle workers and they can't do everything at once. Some of our suppliers are even threatening to refuse our business if we don't get our act together. We've tried to offer solutions for the field service people, but nothing seems to work. Maybe they just don't care."

Mary Shoulton, Chief Financial Officer, "If you can help us with this inventory problem you'll be well worth your salary, and then some! Here we are being competitively crunched for price, delivery, and efficient service and our service inventory costs seem to have gone completely out of control. The total inventory has climbed more than 200% in the last two years while our service revenues have only grown 15%. On top of that, we have had an increase in obsolete material write-off of 80% in that same two-year period. In addition, significant inventory-related costs have come from expediting. Premium freight shipments, such as flying in parts, caused by critical part shortages cost us over $67,000 last year alone. Do you realize that represents almost 20% of our gross profit margin from the service business? With our interest rates, warehousing, and obsolete inventory costs, we recognize a 23% inventory holding cost. Given our huge inventory level, that takes another big bite out of profits. All this suggests to me we need to get control of the situation or we may find ourselves out of business!"

Franklin Knowles, Field Service Supervisor, "Until they hired you, the other production supervisor and I had been in charge of inventory. I hate to discourage you, but it looks like an impossible job. The purchasing people bought a bunch of standard-size bins, and they told us that as soon as we had a week's average part usage for each part to order more—specifically, "enough to fill up the bin." Since most of their lead times were a week or less, it sure made sense. All the records were kept on computer, therefore the computer could be programmed to tell us when we had only the week's supply. It made great sense to me, but something kept going wrong. First, field service technicians seemed to frequently grab parts without filling out a transaction. That made our records go to pot. As a matter of fact, we had a complete physical inventory a couple of months ago, and it showed our records to be less than 30% accurate! I suspect our records are almost that bad again, and we don't have another physical inventory scheduled for another nine months.

"Second, with our records so bad the field service technicians can never tell if we really have the parts or not. Several of them have started to take large quantities of critical parts and keeping their own inventory. When it comes time to replace their own "private stock," they take a bunch more. That has made the demand on the central inventory appear very erratic. One day we have plenty, and the next day we're out! You can imagine how happy purchasing is when the first time they see a purchase order it is requesting an immediate urgent shipment. We've made a policy that the technicians are only supposed to have a few specifically authorized parts with them, but I'm sure many of the technicians are violating that policy big-time."

Quentin Bates, Field Service Technician, "Something is drastically wrong with our inventory, and it's driving me and the other techs crazy. We're not supposed to keep much inventory with us, only a few commonly used parts. If we have a field problem requiring a part, we're supposed to be getting it from the central inventory. Problem is, much of the time it's not there. We have to take time to pressure purchasing for it, and then have to try to calm our customers while we wait for delivery. In the meantime, the customers' systems are often unusable, and they're losing business. It doesn't take too long before they're really mad at us. I guess the people at purchasing don't care, since we have to take all the heat. Lately I've been taking and keeping a bunch of parts I'm not really supposed to have in my inventory, and I know the other field technicians do also. That's saved us a few times, but the situation seems to be getting worse."

Now that Rosa had some sense as to the nature of the problems, she needed to start developing solutions—and it appeared that it was important to come up with good solutions fast! The first thing she tried to do was take a couple of part numbers at random and see if she could improve on the ordering approach.

The first number she selected was the A233 circuit board. The average weekly usage was 32, with a standard deviation of 47. The lead time was given as one week. The board cost $18, and the cost to place an order was given as $16. The quantity ordered to fill the bin was usually 64. The second number was the P656 power supply. It cost $35, but since the supplier only required a fax to order the cost was only $2 per order. Even with the fax the delivery lead time was two weeks. The average weekly demand for the power supply was 120 with a standard deviation of 14 units. They typically ordered 350 units at a time. Recently the supplier for the circuit board hinted that they might be able to give Carl's a price break of $2 per board if Carl would order 200 or more at a time.

CASE ANALYSIS

1. Using the data on the two part numbers given, provide a comprehensive evaluation of their ordering policies. Compare the present annual average cost with the cost of using a system such as EOQ, and discuss any other order policies as appropriate.
2. Should Carl's pursue the price break? Why or why not?
3. What do you think the sources of the other problems are? Be specific and analyze as completely as possible.
4. Develop a comprehensive plan to help Rosa put the inventory back in control.

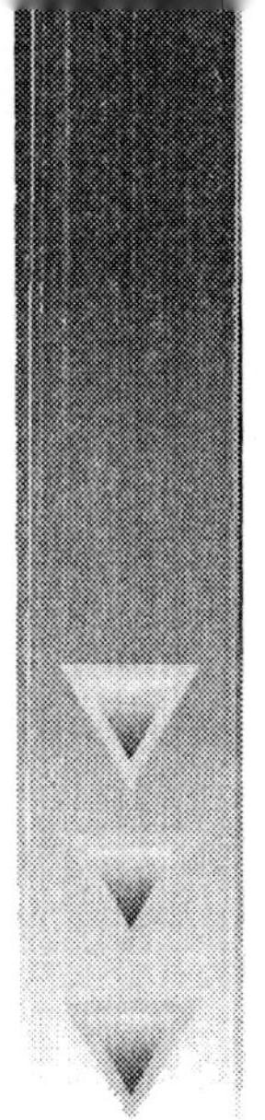

CASE 20

THE HAPPY WANDERER MOTOR HOME COMPANY

Fleur always enjoyed her job as a planner, and in the two years that she had held this job since graduation from college she embraced change as a fact of doing business. But this time, it seemed that management was asking too much. At the recent management-finance meeting it was announced that the company was to go through an exercise in "right-sizing" across all departments. The materials manager, Bert Snow, knew that Human Resources would be assessing the workloads of all departments, and he had a strong suspicion that this exercise would conclude with a required reduction of two staff from materials management. With his usual candor and sincerity, Bert quickly announced this to his staff. "Better you hear this from me rather than letting it get out of hand in the rumor mill," was his comment.

THE CURRENT SITUATION

Fleur, who had the ability to step back and look at the problem from a company perspective, wondered how "right-sizing" could be done since she felt that everyone was already stretched to the limit. She also wondered about management's apparently arbitrary decision to reduce staff in her department, since the cost of materials exceeded 80% of the cost of the finished product. She quickly pulled a copy of the department organization chart that had been used in the last team building session and reminded herself to think of how to "get the work out of the system." She knew that there were no slackers in her department, but she decided to make a mental note of what she saw everyone doing as part of their job duties.

As the planner responsible for the interior trim components of the motor homes, Fleur's daily tasks included reviewing requisitions prior to their release (to ensure all requisitions have real needs), releasing parts from blanket purchase orders, rescheduling and cancelling orders as required, and monitoring daily inventory transactions for unexpected demands. The items she controlled included everything, such as carpet on the floor, entertainment system components, bathroom fixtures and, yes, the kitchen sink. She was gradually learning more of the technical aspects of these items but left a lot of that up to the buyers. She felt that any increase in her duties would only cause her to make more mistakes, resulting in more cost to the company.

The other planners had similar duties and readily shared information regarding any changes that would affect the demand for Fleur's materials. If a shipment was delayed or an engineering change affected an order, the other planners were quick to alert Fleur to avoid problems in her area. Lately the buyers and planners were even getting together every morning for about a half hour to discuss current orders and any changes. These meetings were very productive since the shortage of one component could affect many customer orders. The planner responsible for structure also looked after all the plumbing and electrical work. Fleur couldn't see how this work could be split-up and handled by any fewer people.

John, one of the buyers, is dedicated to the chassis and drivetrain because it is the single most expensive component. The position requires constant communication with Ford Motor Company to ensure the chassis arrive at the right time with the correct configuration of engine, frame, tires, and brakes. Many variations are possible depending on the final model being built. It wasn't the complexity of this job that was important but rather the cost of making up for any errors or last-minute changes. John was constantly on call to handle communications between Ford, marketing, and cost accounting. The other two buyers were split between local and distant suppliers with some components being sourced in Japan. The buyers have a sound technical knowledge of their products and work closely with the engineering department. Again, Fleur could not see how to cut any of the activities the buyers performed as part of "right-sizing."

The two expediters are known for their ability to find missing materials in the plant and beg favors from the shop supervisors. Expediters have to be high-energy people, willing to chase down problems and get the orders completed. They also require good memories to be able to replace components that they may have "borrowed" from one order to get a rush job through. They report directly to Fleur and the structure planner, but also get requests to help the buyers. The buyers sometimes asked them to work directly with suppliers to continuously track materials from the supplier into the factory. Fleur worried that the expediters might be the first employees to go. It was common knowledge that if things were done right then expediters would not be needed. What really scared her was that if they took away her expediter this would probably mean that she would have to spend some of her own time going to the shop floor, calling suppliers, and generally letting more important things slide.

Two clerks are responsible for office support to Mr. Snow. Their duties involve filing and postage of purchase orders. One clerk looks after customs documentation for imported goods. These people were constantly busy and really helped to keep the entire office operating in an orderly fashion. Fleur knew that if one of them went then she would miss the security of knowing that files were in order and that foreign-source goods would arrive on time.

Bert, probably the busiest person in the department, sets the high-energy pace for everyone who reports to him. As materials manager he has department responsibility for setting budgets and ensuring open communications between his staff and other departments. He reports to the Executive Vice-President, who keeps him pretty busy producing reports and attending meetings. In addition to these responsibilities, he also conducts most of the negotiations with major suppliers. This is done on an ongoing basis and consumes about 25% of his time. He always includes the respective buyers in negotiations and feels that this improves the long-term relations of buyers and suppliers plus grooms the buyers for future management responsibilities.

Fleur wondered how Bert would approach this difficult decision of staff reduction. A department meeting was planned for the afternoon, and she wanted to make a positive contribution rather than appear a complainer in the face of progressive change.

CASE ANALYSIS

1. What steps should Bert take to reassign job responsibilities within his department?
2. How would the use of Buyer-Planners reduce the amount of work being done?
3. Draw a proposed organization chart for the department.

The Happy Wanderer Motor Home Company Organization Chart—Materials Management Department

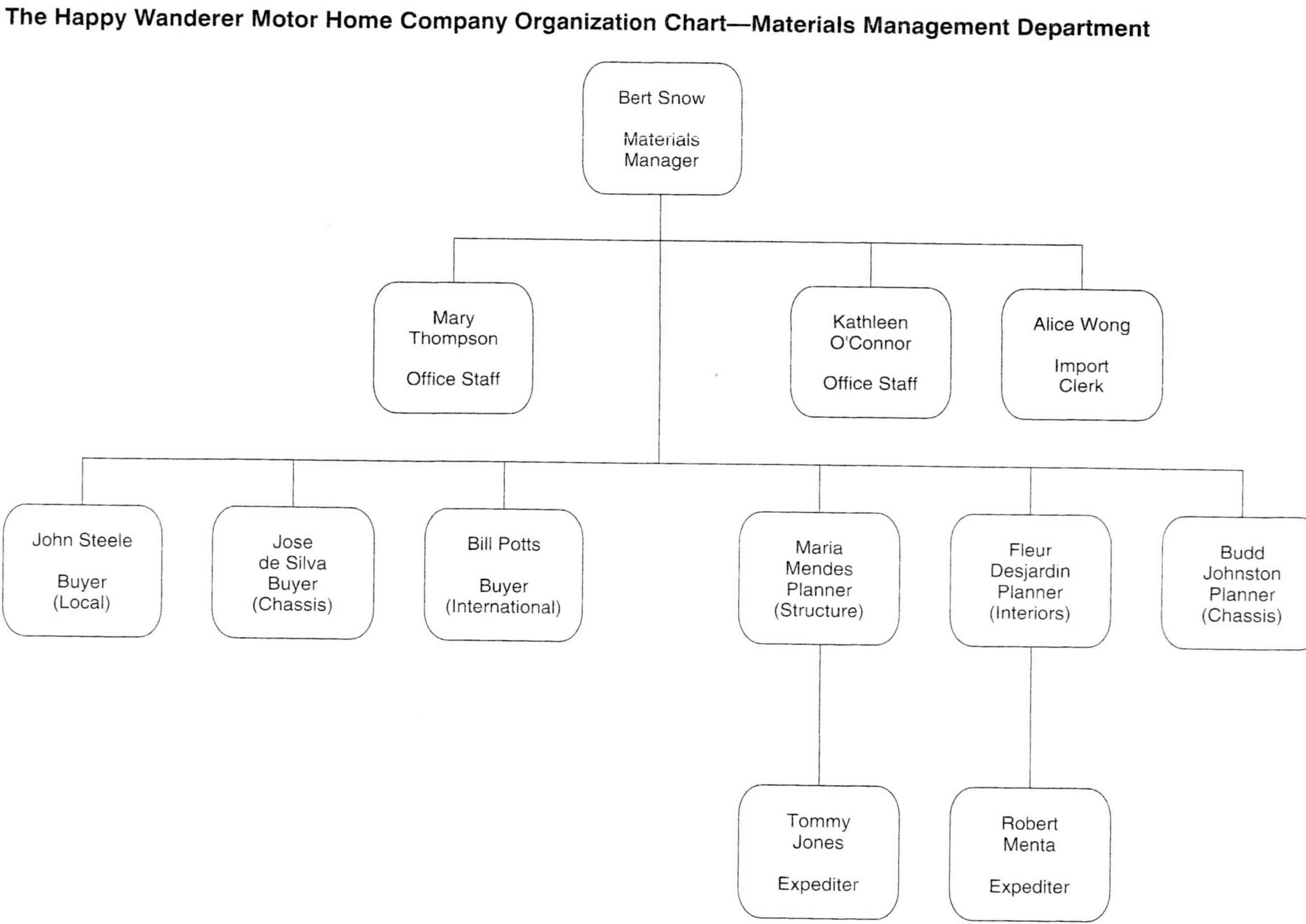

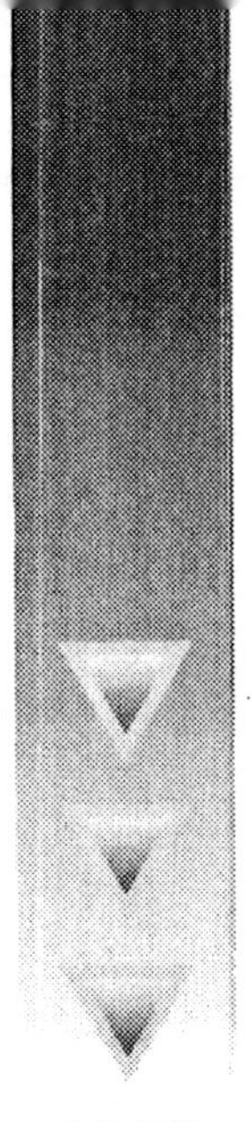

CASE 21

MIGHTY MOUNTAIN BIKES, INC.

Mighty Mountain Bikes is a small company that specializes in making mountain bikes. The two owners were avid mountain cyclists when they started the business, and they built it based on their knowledge of mountain bikers' needs. The company has grown, and today they ship to specialty shops all over the world.

MMB makes several different bikes distinguished by frame size, wheel size, number of gears, and trim packages. The packages include the paraphernalia mountain bikers need—water bottles, special tires, handle grips, and so on. MMB makes their own frames and buys other components. They assemble bikes to order using mostly standard parts. Generally they can assemble the bike the customer wants from a selection of three wheel sizes, three frame sizes, three gearshifts, and two basic trims—standard and deluxe.

The business has been running for a number of years and is well established, but this is mainly due to the experience and energy of the two founders. Since there is every indication that the business will continue to grow, the owners feel that they cannot cope in their present manner.

Forecasting and production control have always been a problem that is increasing as the business grows. Recently, they hired you as the new materials manager to take over these responsibilities. They have kept some records and the following data have been gathered for sales of all bikes over the past five years. Year 5 is last year.

Historical Sales of Mighty Mountain Bikes					
	Sales in units				
Month	Year				
	1	2	3	4	5
January	600	640	720	750	835
February	700	720	780	800	850
March	1100	1130	1260	1340	1390
April	1350	1360	1460	1475	1535
May	1535	1620	1760	1830	1960
June	1550	1635	1750	1840	1950
July	1510	1585	1670	1740	1845
August	1475	1560	1660	1750	1855
September	1200	1245	1310	1345	1415
October	1150	1175	1200	1220	1245
November	800	825	850	875	900
December	1200	1225	1250	1250	1275
Total	14170	14720	15670	16215	17055

Of the 17,055 bikes sold last year the records show the following breakdown of parts used.

Part	**Part Number**	**Usage**
Frame	101	5,117
	102	7,675
	103	4,264
Wheel	201	2,558
	202	6,822
	203	7,675
Gear	301	1,706
	302	7,675
	303	7,675
Trim	401	6,822
	402	10,233

Since there are three frames, three wheels, three gears, and two trims, Mighty Mountain Bikes makes a total of 54 different bikes.

Case Analysis

Develop a forecast for the demand for bikes for year 6. Bear in mind that you will use the forecast to develop a production plan and a master production schedule. What should you forecast for each? What forecasting techniques might be appropriate to use? Be prepared to justify your choice of technique.

Having developed your forecast, develop a level production plan for the 12 months of year 6. MMB operates for 50 weeks of the year, closing down for two weeks in July. For year 6, the months have the following number of weeks.

Month	Number of Weeks	Month	Number of Weeks
January	5	July	2
February	4	August	5
March	4	September	4
April	4	October	5
May	5	November	4
June	4	December	4

The year-end inventories in units for the three frames are as follows. MMB wishes to reduce that inventory to 800 units by the end of the year.

Frame 101	300
Frame 102	450
Frame 103	250
Total	1000

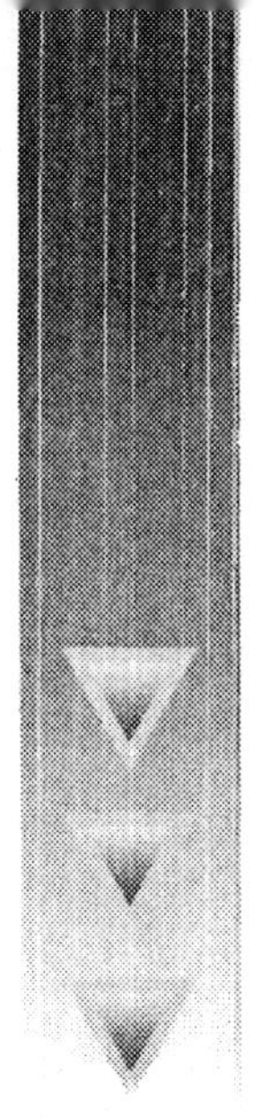

CASE 22

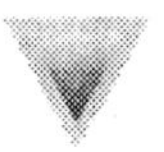

NORTHCUTT BIKES II: THE FORECASTING PROBLEM

INTRODUCTION

Jan Northcutt, owner of Northcutt Bikes, started business in 1975 when she noticed the quality of bikes she purchased for sale in her Raleigh bike shop declining while the prices went up. She also found it more difficult to obtain the features she wanted on ordered bikes without waiting for months. Her frustration turned to a determination to build her own bikes to her particular customer specification.

She began by buying all the necessary parts (frames, seats, tires, etc.) and assembling them in a rented garage using two helpers. As the word spread about her shop's responsiveness to options, delivery, and quality, however, the individual customer base grew to include other bike shops in the area. As her business grew and demanded more of her attention, she soon found it necessary to sell the bike shop itself and concentrate on the production of bikes from a fairly large leased factory space.

As the business continued to grow, she backward integrated more and more processes into her operation, so that now she purchases less than 50% of the component value of the manufactured bikes. This not only improves her control of production quality but also helps her control the costs of production and makes the final product more cost attractive to her customers.

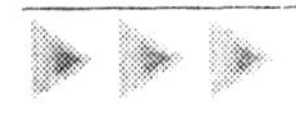

THE CURRENT SITUATION

Jan considers herself a "hands-on" manager, and has typically used her intuition and her knowledge of the market to anticipate production needs. Since one of her founding principles was rapid and reliable delivery to customer specification, she felt she needed to begin production of the basic parts for a particular style of bike well in advance of demand. In that way she could have the basic frame, wheels, and standard accessories started in production prior to the recognition of actual demand, leaving only the optional "add-ons" to assemble once the order came in. Her turnaround time for an order of less than half of the industry average is considered a major strategic advantage, and she feels it is vital for her to maintain or even improve on response time if she is to maintain her successful operation.

As the customer base has grown, however, the number of customers Jan knows has significantly shrunk as a percentage of the total customer base for Northcutt Bikes, and many of these new customers are expecting or even demanding very short response times; since that is what attracted them to Northcutt Bikes in the first place. This condition, in addition to the volatility of overall demand, has put a strain on capacity planning. She finds that at times there is a lot of idle time (adding significantly to costs), while at other times the demand exceeds capacity and hurts customer response time. The production facility has therefore turned to trying to project demand for certain models and actually building a finished goods inventory of those models. This has not proven to be too satisfactory, since it has actually hurt costs and some response times. Reasons include:

- The finished goods inventory is often not the "right" inventory, meaning shortages for some goods and excessive inventory of others. This condition both hurts responsiveness and increases inventory costs.
- Often to help maintain responsiveness, inventory is withdrawn from finished goods and reworked, adding to product cost.
- Reworking inventory uses valuable capacity reserved for customer orders, again resulting in poorer response times and/or increased costs due to expediting. Existing production orders and rework orders are both competing for vital equipment and resources during times of high demand, and scheduling has become a nightmare.

The inventory problem has grown to the point that additional storage space is needed, and that is a cost that Jan would like to avoid if possible.

Another problem Jan faces is the volatility of demand for bikes. Because she is worried about unproductive idle time and yet does not wish to lay off her workers during times of low demand, she has allowed them to continue to steadily work and build finished goods. This makes the problem of building the "right" finished goods even more important, especially given the tight availability of storage space.

PAST DEMAND

The following list shows the monthly demand for one major product line—the standard 26-inch, 10-speed street bike. While being only one product, it is representative of most of the major product lines currently being produced by Northcutt Bikes. If Jan can find a way to use this data to more constructively understand her demand, she feels she can probably use the same methodologies to project demand for other major product families. Such knowledge can allow her, she feels, to plan more effectively and continue to be responsive while still controlling costs.

Demand per Month	1996	1997	1998	1999
January	437	712	613	701
February	605	732	984	1291
March	722	829	812	1162
April	893	992	1218	1088
May	901	1148	1187	1497
June	1311	1552	1430	1781
July	1055	927	1392	1843
August	975	1284	1481	839
September	822	1118	940	1273
October	893	737	994	912
November	599	983	807	996
December	608	872	527	792

CASE ANALYSIS

1. Plot the data and describe what you see. What does it mean and how would you use the information from the plot to help you develop a forecast?
2. Use at least two different methodologies to develop as accurate a forecast for the demand as possible. Use each of those methods to project the demand for the next four months.
3. Which method from question 2 is better? How do you know that?
4. How, if at all, could we use Jan's knowledge of the market to improve the forecast? Would it be better to forecast in quarterly increments instead of monthly? Why or why not?
5. Are there other possible approaches that might improve Jan's operation and situation? What would they be and how could they help?
6. Has Jan's operation grown too big? Why or why not? What would you suggest she do? What additional information would you suggest she look for to help her situation?

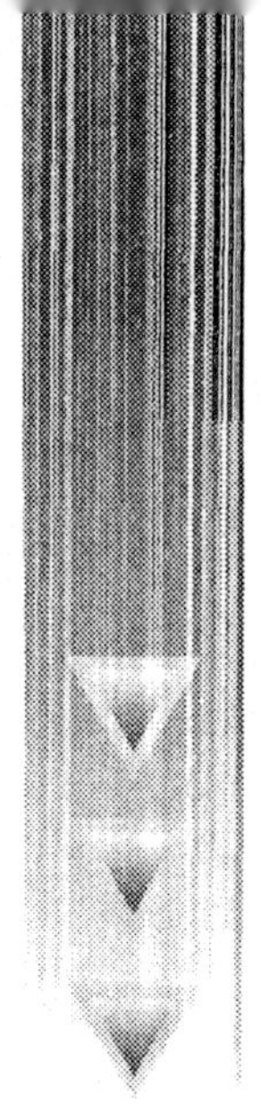

CASE 23

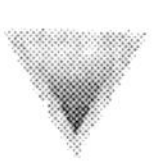

MELROSE PRODUCTS

Jim Harding was not in a good mood. He worked his way through the "ranks" when "supervisors did supervision and workers did what they were told to do." He was now faced with the fact that the new president of Melrose Products was one of those "touchy-feely" types pushing for self-directed work teams. As the manufacturing manager, Jim was ultimately responsible not only to meet production needs, but also to do so in the most efficient and cost-effective manner possible. To him, that meant specific allocation of work. It had always worked that way and he saw nothing new to tell him it shouldn't continue to do so.

Part of the problem, Jim realized, was that the business environment was changing. Changes in the product design were becoming more frequent and the customers were expecting more service. While they were still sensitive to price (the competition had not disappeared), they wanted quick delivery, high quality, and the product designed more specifically to their need. To Jim, that meant putting more pressure on those "lazy, pampered engineers" to make better designs as well as additional pressure on those "bums on the factory floor" to meet production needs. With better designs, he could more easily allocate the work to his workforce to meet the customer demands. He felt he had truly kept up with the times—the customer was king. The fact that the customer expected more meant little more than how to get them what they wanted from production. It was merely a case of making sure everyone delivered on the job the way they were supposed to.

While Mr. Melrose had avoided the need to become a public company and had managed to keep unions out, he still apparently had gone soft, at least according to Jim. He had recently appointed Cindy Lopez as the new president, passing over Jim.

She not only had an MBA (Jim had always thought the real business learning was done "on the firing line"), but she had never even been a supervisor. She had come from, of all places, the Human Resources Department! What had they ever done for him other than send him a bunch of worthless people. Some of those people had, in his mind, no chance of ever becoming useful. As far as he was concerned, the only real value of a Human Resources Department was to keep those government idiot bureaucrats off their backs. All those ridiculous regulations!

So now Jim was in the position to try to "change with the times," as Cindy had said. She wanted to gradually move the company toward flexible self-directed work teams. Jim, of course, felt that all the workers really wanted was to get their paycheck and get drunk on Friday and Saturday nights, and could care less about having any say in the product or the customer. How was he ever going to get anything done with someone so naive in charge?

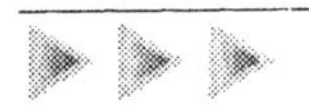

THE CURRENT SITUATION

Cindy had suggested that Jim start the process of changing to teams by looking at the K-line. The K-line of product was a fairly standard product that had recently undergone heavy competitive pressure in the form of delivery speed and design enhancements. Melrose had been gradually losing market share in the K-line. Jim had responded (before the naming of Cindy as president) by putting additional pressure on workers to be more efficient and to reduce their task times. As Jim said, "There's always some slack time we can squeeze out of any process if we really put our minds to it."

They are using carefully developed time standards, much as Jim learned in his Industrial Engineering courses. He feels they are quite good, including a liberal 10% allowance. Since the K-line is a fairly standard product, Jim uses the time standard not only to develop cost figures for labor, but also to use those cost figures to allocate overhead.

There are currently seven labor tasks to make one of the K-line products.

Task	Standard Time (Min.)	Estimated Labor Cost/Minute
1	7.5	$0.24
2	2.3	$0.22
3	4.7	$0.28
4	5.1	$0.29
5	17.8	$0.26
6	19.1	$0.18
7	8.4	$0.25

The overhead allocation is currently at 230% of direct labor. Material costs are $9.35 per unit. They currently have enough labor to produce 20 of the K-line per shift. Each shift has one supervisor costing about $24 per hour.

From this information, Jim was being asked to develop teams, and without a supervisor. From his standpoint, the effort was doomed to failure. Jim, however, always considered himself a company man and would do what he could to make it happen.

CASE ANALYSIS

1. What is the standard cost of the K-line product?
2. What steps (specifically) would you undertake to make the self-directed teams? How (specifically) would you deal with the cost and time standard issues?
3. Do you agree with Cindy? Do you agree with Jim? Is there some alternative approach that might be better in this situation? Explain.
4. What would you do with the supervisor in this situation? Be specific in your approach.

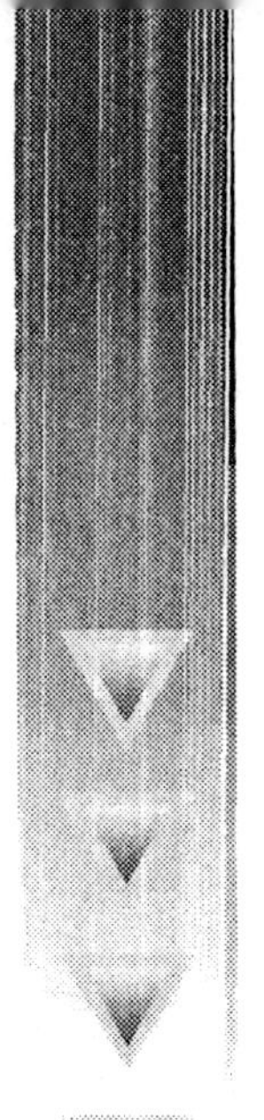

CASE 24

THE COMPETITIVE BIDDING TRAP

Frank was starting to glaze over in class. He had been attending night classes for some time now and he thought that he would be immune to the bad habits of his undergraduate years. Work was taking a toll on him, but this time it was something else. His professor had used the phrase "The Competitive Bidding Trap," and Frank was thrown back to this past week's activities. In truth, it made him feel like he was in a trap. Quickly Frank snapped out of his temporary lack of attention and wrote down the five prerequisites to competitive bidding. This seemed to get him back on track for the class.

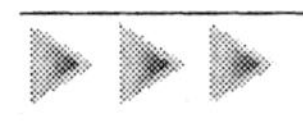

THE FIVE PREREQUISITES FOR COMPETITIVE BIDDING

1. The dollar value of the purchase must be worth the effort of bidding for both the supplier and the buyer.
2. The specifications for the item must be clear to both the supplier and the buyer.
3. There must be an adequate number of suppliers.
4. The suppliers must be both technically capable of supplying the product and interested in securing the contract.
5. There must be sufficient time for suppliers to generate quotations and for buyers to analyze them. Thirty days is considered a minimum.

Frank had been working for two years now in the buying department for Pennsylvania Surgical. They produced a line of low-cost single-use surgical instruments for hospitals, doctors, and dentists. He had worked his way up from the position of expediter, which he had taken just out of college, and was now a full-fledged buyer with responsibility for about $6 million per year of raw materials and supplies for manufacturing. With a college diploma in materials management it was a junior position for him to start at, but Pennsylvania Surgical was a fast growing company located close to his hometown. Things had worked out well for him and with his determination to continue toward certification with the National Association of Purchasing Managers (NAPM) he felt that he could one day be a senior buyer or even better. He was known to be very conscientious with his work and constantly searched for ways to save the company money.

THE CURRENT SITUATION

Thirty days ago Frank had sent a request for quotations to a number of potential suppliers of a special instrument lubricant that Pennsylvania Surgical used to treat all their surgical instruments prior to sterilization and packaging. Pennsylvania Surgical used an autoclave to sterilize their instruments and this can be very corrosive, even for stainless steel. To protect and lubricate the instruments they soaked them for a few minutes in a water-oil mixture. During sterilization the water was boiled off the surface of the instruments and a very fine layer of oil remained.

It seemed that for a small number of companies, mixing oil and water was easy to do, since Frank had received four quotes to provide approximately 10,000 gallons of the mixture per year. Earlier this week he had opened the quotes on the specified date and found that once again Minnesota Mixers had won the bidding process by a significant margin. The quotes FOB Pennsylvania Surgical were as follows:

Company	Price per Gallon ($)
Detroit Lubricants	27.04
Minnesota Mixers	25.94
Blend-it Solutions	28.40
New York Chemical	29.20

At work the next day Frank was talking to Phil, a friend from engineering, and he asked the question, "Just exactly how do you get oil and water to mix?"

His friend smiled and said, "Oh, are you getting some more instrument milk? That's what we call the stuff used to protect the instruments during sterilization. It looks like milk. But, don't try to drink it."

Frank felt that it was now okay to talk in general about the bids. All the bids had been opened and company policy would not allow any further bids to be accepted. "Yes, I just opened four legitimate bids and Minnesota Mixers was lowest again for the fourth year in a row. Aren't the other companies interested in our business?"

Phil explained that he had worked on the instrument lubricant when they first piloted an in-house experiment on reducing corrosion within the package. The key component was a high-shear pump, also known as a homogenizer. Also, not keeping the stuff around forever, but getting freshly mixed lubricant every month prevented separation of the oil and water. The company had decided to have the lubricant made by a specialty supplier due to the cost of equipment and the need for production space. Phil was just getting up to leave the table but added, "The equipment to make that stuff in the volume we need costs about $50,000 and I'll bet that we requested dedicated equipment that could not be used to mix other non-medical product."

When Frank returned to his office he went to the commodity file on instrument lubricant and found the original quotes for the product. Four years ago Minnesota Mixers had won the bidding with a quote of $24.57. The next closest bid was $24.63. This seemed reasonable with a factor for slight inflation since then. He did not see any information on setup costs for the production and asked the senior buyer, Gerard Schtall, if he could remember anything about the original bids.

"Yes, we did not separate the setup and per unit costs of the instrument milk. We felt it was outside our realm of expertise and left this up to the supplier. Are some of the quotes out of line?"

Frank explained that the quotes were all very consistent, perhaps too consistent.

Next he informed the other bidders of their failure to win the bidding again, and then confirmed the price with Minnesota Mixers. The sales rep from Detroit Lubricants was pretty vocal about losing again and the effort that he had put into giving his lowest possible price.

"I'm sorry, but the way that you send out for new quotes every year does not assure us of future business. I have to include a one-year write-off of equipment in my price. We included a setup of $51,500. Perhaps we could renegotiate under different terms. After all, you will probably be using instrument milk for the next few years."

Frank informed him of the policies that he honored regarding competitive bidding and that everyone was treated equally. He did, however, wonder about the one-year term for instrument lubricant contracts. He now wondered if the competitive bidding trap had something to do with all of this.

Case Analysis

1. What condition made competitive bidding not work in this case?
2. Was competitive bidding a good method to use for the first contract for instrument lubricant?
3. What target price should Frank aim for the next time he asks for quotations on instrument lubricant? Should he use a method other than price per gallon?
4. Suggest a pricing strategy for Frank to use next time he asks for quotations.

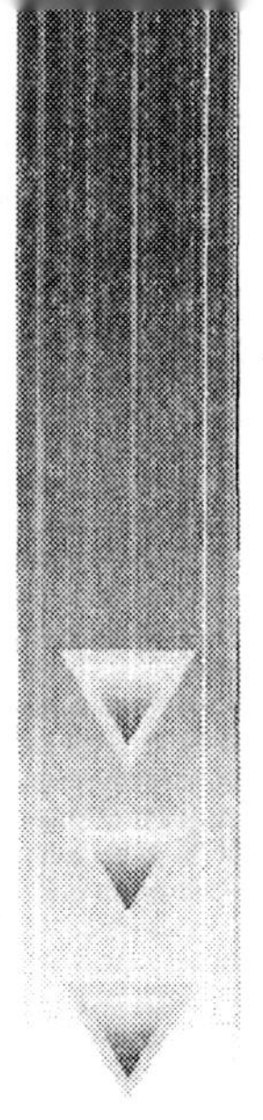

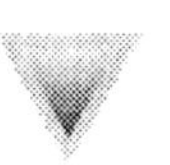

CASE 25

CRANFIELD MANUFACTURING COMPANY

Ravi Gupta finally was hired for the position of his dreams. As the new materials manager for the Cranfield Manufacturing Company, he would be able to put into practice the knowledge he learned in school. Cranfield was a fairly small, privately owned company that, like many similar companies, had evolved from a single entrepreneur (John Cranfield) with a good idea for a new product. Also as is common for such companies, the formality of systems and procedures tended to lag behind their need significantly. That was one of the major reasons Ravi was hired. It was a newly created position since Mr. Cranfield suspected the materials activity of the operation was not as cost-effective as it might be.

One of the first assignments for Ravi was to analyze the materials ordering practices. To do this he started his investigation by discussing present practices with Jaime Hernandez, the general superintendent of the operation. Jaime had moved "up the ranks" to his present position. He was one of Mr. Cranfield's original employees, starting in the company as a machine operator. According to Jaime,

"Most of us here have no formal training in materials management. We've done pretty well, though. We know enough to order in large enough quantities to prevent ordering all the time, yet we also know that ordering too much could lead to storage and spoilage problems, to say nothing of the money that excess inventory ties up. We pretty much use our common sense for ordering production quantities, but tend to rely on the advice of Kanesha (Kanesha Williamson, the head buyer) as to what to order for purchase parts. Basically, when the inventory of a quantity gets low enough, we order more. We do have some idea of the replenishment lead time for each, and have used our experience to set some basic reorder points for most of what we need."

It was obvious the next move was to talk to Kanesha. She had more formal education than most in the facility, but Ravi suspected that she probably could have used a bit more business training than the history degree she had earned from the local university. She had been with Cranfield for five years, however. According to Kanesha,

"I have mostly relied on our supplier to help us set lot sizes and lead times. They've been really great to work with. I spend most of my time working on price and design issues with them. Once we have those figured out, I just ask them what they think a reasonable lot size and lead time should be. Their numbers get us into trouble once in a while, but if we need an emergency shipment most of them can respond if we agree to a high enough premium price. I think most of them use transportation cost as a major factor to set the lot size, and I suppose they know best about those issues."

Ravi suspected that the reason this practice went on without question was the fact that most of the suppliers had pretty good product quality. At least that was one thing he didn't have as an immediate worry.

Ravi learned about MRP and JIT in his classes, and suspected that one would be superior to their current system of using reorder points for production material. He also thought that given the erratic demand patterns and frequent design changes for many of their products, JIT probably would not be the best choice. He therefore did some investigation on the cost and effort to put in a simple MRP system and presented the idea to Mr. Cranfield. Cranfield said he had heard of such systems, but could not justify using one unless it could be shown that the system would save more money than it cost.

Ravi set about getting some data he could analyze to prove the value of such a system. He selected one product that was clearly representative of the overall product mix of Cranfield. The following data were collected:

- *Part number 256-1.* Used to produce part 344, a subassembly for final product A603. There is one part 256-1 required for every 344 subassembly. The lead time to build the 344 from the 256-1 (and other components) is one week.
- *Part number 344.* Used to make final product A603. Product A603 is a final assembly sold directly to customers on an assemble-to-order basis. There are two part 344 required for each A603. The production lead time for A603 is 1 week.
- *Orders for A603.* The customers tended to order the product far in advance, even though the lead-time was rather short. The following orders tended to be fairly firm for the next 26 weeks (half a year):

Week (from present time)	Demand for A603
5	15
7	50
9	120
11	25
15	100
18	50
22	30
23	150
26	40

- *Order information.* The supplier could supply part 256-1 with a two-week lead-time. They were using a present order quantity of 250 units with a reorder point of 45 units (which is the average weekly usage of the part). The cost of placing an order with the supplier was estimated to be $55. The supplier has a policy that once a standard lot size is set, all orders were to be that quantity or any amount more. For example, if they needed 300 units in one week, they could order exactly 300. If they needed 270 units in a week, they could order 270. If, however, they only needed 190 units, they must order the full 250. This policy would stay in effect even if they changed the standard lot size. For example, if they changed the standard order quantity to 150 units, then they could order any amount above 150 but not less than 150. The supplier allowed them to reevaluate and change order quantities no more than once a year.

  They had used the 250 order quantity for the last two years, but were thinking of making it larger. They didn't like the high order cost and they also thought that a larger quantity would minimize the number of times per year they were exposed to a possible stockout. The stockout issue was important to them. They estimated that every time they had a stockout and needed to expedite the order it cost them an additional $300 (in addition to a lot of stress).
- *Inventory cost.* The company accountant estimated that taking all factors into consideration, the holding cost for the 256-1 item was about $1 per unit per week. The actual item cost was $315 per unit. There are currently 80 units in inventory.
- *Lot size for 344 and A603.* Assume they are both produced on a lot-for-lot basis.

CASE ANALYSIS

1. Based on the information given and the current order practices, compute
   a. The projected inventory cost per year
   b. The projected order cost per year

c. The number of times a shortage could cause expediting activity
d. The total cost for the item for the year

HINT: Do NOT use (Q/2)H to find holding cost. That model assumes reasonably constant demand and you're not even close to that in this case. Also assume the pattern of orders for the first 26 weeks of the year continues basically the same for the rest of the year.

2. Compute the EOQ for the data given. Using the EOQ as the new lot size, compute the same data that you found in question 1.
3. Now use the EOQ as the lot size for an MRP system, and compute the same information once again.
4. Using this information and realizing that item 256-1 represented about 0.7% of the total inventory cost of the operation (NOT including item costs), what cost differences would exist annually using the three methods? Given the cost of the MRP system is estimated to be $1,500,000 to implement in this operation, how long would it take to pay for itself?
5. Is there anything about the information in the case that you would recommend Ravi to investigate further before he considers moving ahead with any changes? What changes in management policies should he consider, for example? Specifically address any practices discussed in the case that may be inappropriate for MRP. Try to put any changes into a brief implementation approach.

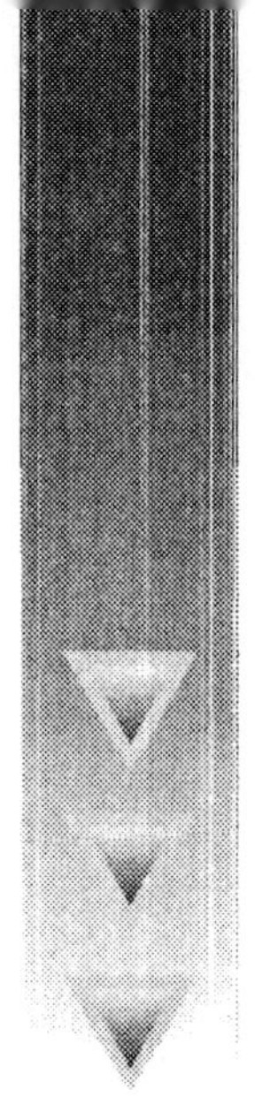

CASE 26

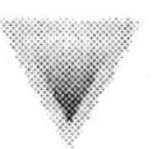

MOTO-TRACTOR COMPANY

The Moto-Tractor Company makes a line of heavy-duty garden equipment for the retail market. Their products are extremely durable and expected to have a serviceable life span of 20 years or more. Not only is the product very rugged due to sound engineering design, but also it is designed to be easily repaired. The company has been in business since 1953 and their products have increased from one basic model to over 20 variations.

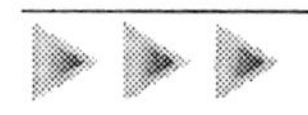

THE CURRENT SITUATION

Tony Carravagi has been the materials manager for about three years now and is getting lots of pressure from the vice-president to reduce the costs of inventory. This has always been a frustration for Tony, since the company has some well-established traditions that he just can't seem to overcome. To reduce inventory costs Tony knows that real improvement will only come from changing some of the aspects of how they do business, not just by fiddling with some factors in the EOQ formula they are presently using.

Engineering has a lot of pull at management meetings. Tony acknowledged the support they get because the product is very rugged in design. However, anytime Tony challenged a new design or suggested a material substitution all he got were blank stares and closed minds. At the last management-finance meeting, engineering was reporting on the progress of design work for the next product upgrade. Tony was scanning the schedule of design work and noted the listing of a carburetor mounting bracket, but didn't notice the design of a new carburetor. He decided not to ask the obvious question right away but planned to visit his one ally, Harvinder Ramdial in engineering.

After the meeting Tony returned to his office and grabbed a quote on a new supplier of fasteners as well as the inventory listings for carburetor mounting brackets and headed for engineering. Harv greeted him with the customary "Are you lost?", since few braved the offices of the engineering department, even with a cover story of investigating a new supplier.

Tony showed Harv the listing of mounting brackets and said, "At this morning's meeting I noticed that the designers are not working on a new carburetor but are working on a new mounting bracket. Pardon my business background, but why is this?"

"They've moved the cylinder-side bolt spacing again" said Harv. "Here, let me show you." Harv who was a whiz at the new CAD software, turned to his computer and started punching the keys while explaining, "One of my first jobs here was to enter those parts into the new system from the old drawings. Watch this. What I am going to show you is the 106 series of mounting brackets, in sequence."

Harv proceeded to quickly show drawings of four different mounting brackets which Tony thought all looked the same. "Okay, so what am I supposed to notice?"

"Look again at the hole spacing on the cylinder side" said Harv. Sure enough, this time Tony noticed that everything else in the drawings looked identical while the mounting holes moved around and changed size.

Harv explained, "When I was entering the old blueprints into the CAD system I noticed the similarities in all the mounting brackets as well as the differences. Being wise in the use of my resources I copied each file and only made the necessary changes. You should have seen the look on old Sid's face when I completed the drawings in record time. Sid has been hand-drawing these for years and has always started from scratch. I went on to suggest that an oversized slot instead of a located hole would make all the brackets interchangeable, but they shot my idea down since it would increase the product cost by about 2%. The slot would work, but it's easier to drill a hole than to mill a slot. Besides, to replace the eight items you have in your list would require two universal brackets. A two-hole for the 106 series and a four-hole for the 108 series."

Tony returned to his office and started gathering some information. He printed out the following list and also noted the assumptions used for this product: carrying cost of 35%, safety stock of one-month average demand, and a setup cost per batch of $100. Since there were a lot of calculations and he wanted to test the effect of some management variables, Tony decided to write a small spreadsheet to check the costs of inventory and production of the present brackets and the two proposed brackets.

Product Code	Annual Volume	Cost Per Unit
106877	13,242	$10.95
106879	9,757	$10.15
106983	10,033	$9.97
106996	7,689	$8.78

Product Code	Annual Volume	Cost Per Unit (*continued*)
108105	27,359	$15.42
108234	21,123	$14.87
108542	19,086	$13.29
108569	23,097	$17.76

CASE ANALYSIS

1. Can an argument be made to change the design of the brackets? Can the present policy on safety stock be challenged to make this a profitable decision? Who would be involved in this type of decision?
2. What positive effects would occur with forecasting, safety stock, customer service, or the cost of each setup by changing to the new bracket?
3. Would there be additional benefits to the retail distributors by changing?

CASE 27

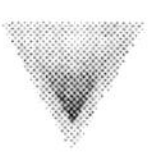

NORTHCUTT BIKES III: THE SERVICE DEPARTMENT

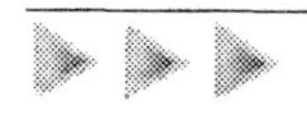

INTRODUCTION

Several years ago, Jan Northcutt, owner and CEO of Northcutt Bikes, saw the need to organize a separate department to deal with service parts for the bikes her company made. Since the competitive strength of her company was developed around customer responsiveness and flexibility, she felt a separate department focused exclusively on aftermarket service was critical toward meeting that mission.

When she established the department, she named Ann Hill, one of her best clerical workers at the time, as manager of the department. At first, it occupied only a corner of the production warehouse, but now it has grown to occupy its own 100,000 square foot warehouse. The service business has also grown significantly, in that it now represents over 15% of the total revenue of Northcutt Bikes. The exclusive mission of the service division is to provide parts (tires, seats, chains, etc.) to the many retail businesses that sell and service Northcutt Bikes.

While Ann has turned out to be a very effective manager (and now has the title of Director of Aftermarket Service), she still lacks a basic understanding of materials management. To help her develop a more effective materials management program, she hired Mike Alexander, a recent master's graduate of a business management program, to fill the newly created position of Materials Manager of Aftermarket Service.

THE CURRENT SITUATION

During the interview process, Mike got an impression that there was a lot of opportunity for improvement. It was only after he selected his starting date and requested some information to be available that he began to see the full extent of the challenges that lay ahead. His first day on the job really opened his eyes.

One of the first items he requested was a status report on inventory history and shipped orders. In response the following note was on his desk the first day from the warehouse supervisor, Art Demming,

"We could not compile the history you requested, as we keep no such records. There's just too much stuff in here to keep a close eye on it all. Rest assured, however, that we think the inventory positions on file are accurate, since we just completed our physical inventory last week. I was able to track down a demand history for a couple of our items, and that is attached to this memo."

When Mike learned this, he decided to investigate further. Although the records were indeed difficult to track down and compile, by the end of his second week he had obtained a fairly good picture of the situation based on an investigation of 100 parts selected at random. He learned, for example, that although there was an average of more than 70 days worth of inventory, the fill rate for customer orders was less than 80%. The remaining orders were back ordered. Unfortunately, many customers viewed many service parts as somewhat generic, and would take their business elsewhere when parts were not available from Northcutt Bikes. What really hurt was those businesses that sometimes canceled their entire order for parts and placed it with another parts supplier. Customers canceled a back ordered item about 15% of the time, and about 5% of the time they would cancel their entire order. Sometimes those customers were lost for good. As the average customer order represented a revenue of about $500, these cancellations had a fairly large impact.

The obvious conclusion is that while there is plenty of inventory overall, the timing and quantities are misplaced. Increasing the inventory did not appear to be the answer, not only because of the large amount already being held, but also because the warehouse space (built less than two years ago) had increased from being 45% utilized just after they moved in to its present utilization of over 95%.

Mike decided to start his analysis and development of solutions on the two items for which Art had already provided a demand history. He felt that if he could analyze and correct any problems with those two parts, he could expand the analysis to most of the others. The two items on which he had history and concentrated his initial analysis were the FB378 Fender Bracket and the GS131 Gear Sprocket. The FB378 is purchased by Northcutt Bikes from a Brazilian source. The lead time has remained fairly constant at three weeks, and the estimated cost of a purchase order for these parts is given at $35.00 per order. Currently Northcutt Bikes uses an order lot size of 120 for the FB378. Northcutt Bikes pays $5.59 apiece for this part. The nature of this product is such that Northcutt Bikes is not offered quantity discounts.

The GS131 part, on the other hand, is a newer product only recently being offered. It is produced for Northcutt Bikes by a machine shop in Nashville, Tennessee, which gives Northcutt Bikes a fairly reliable six-week lead time. The cost of placing an order with the machine shop is only about $15.00, and currently Northcutt Bikes orders 850 at a time. Northcutt Bikes buys the item for $12.85 each. As in the case of FB378, no quantity discounts are offered.

The service division can currently borrow funds at an interest rate of 9%, and is achieving an annual return on assets of 19%. The new warehouse was built two years ago at a cost of $1.8 million, and requires annual average expenditures of $160,000 for utilities, $76,000 for insurance, and $85,000 for maintenance. There are 15 full-time employees at the warehouse, making an average wage of $20 per hour, including benefits. All this translates to what they estimate is a holding charge of 24% per year.

While some of the customers pick up their orders from the Northcutt Bikes Service Warehouse directly, more than 90% have them delivered. Northcutt Bikes contracts with a national delivery company to deliver the orders. The average delivery cost is $12, which Northcutt Bikes adds to the customer's bill.

The supplier for the FB378 Fender Bracket has recently approached Northcutt Bikes with an offer to sell the FB378 at a quantity discount. If Northcutt Bikes bought 300 to 499 per order, their cost would be $5.55. Ordering 500 or more at a time would allow them to cut the cost to $5.50. To this point, Northcutt Bikes has been reluctant to take advantage of the offer because of the cost of inventory and the possible impact on the already strained warehouse space.

The following is the demand information that Art had given to Mike for the FB378 Fender Bracket and the GS131 Gear Sprocket:

	FB 378 Fender Bracket		GS 131 Gear Sprocket	
Week	**Forecast**	**Actual demand**	**Forecast**	**Actual demand**
1	30	34	47	50
2	32	44	56	51
3	35	33	53	46
4	34	39	49	55
5	35	48	51	53
6	38	30	54	51
7	36	26	52	60
8	33	45	55	48
9	37	33	52	53
10	37	30	52	50
11	36	47	50	46
12	37	40	49	55
13	38	31	52	51

(continued)

	FB 378 Fender Bracket		GS 131 Gear Sprocket	
Week	Forecast	Actual demand	Forecast	Actual demand
14	36	38	52	58
15	36	32	55	51
16	35	49	54	44
17	37	24	52	57
18	35	41	53	59
19	37	34	53	46
20	36	24	52	62
21	34	52	53	51
22	36	41	53	60
23	37	30	54	46
24	36	37	53	58
25	36	31	54	42
26	35	45	53	57
27	36		53	

Mike realized he also needed input from Ann as to her perspective on the business. She indicated that she felt strongly that with better management they should be able to use the existing warehouse for years to come, even with the anticipated growth in business. Currently, however, she views the situation as a crisis because, "... we're bursting at the seams with inventory. It's costing us a lot of profit, yet our service level is very poor at less than 80%. I'd like to see us maintain a 97% or better service level without back orders, yet we need to be able to do that with a net reduction in total inventory. What do you think, Mike, can we do better?"

CASE ANALYSIS

1. Use the available information to evaluate Northcutt's inventory management methods. Examine alternative approaches and recommend one. Be sure to include your analysis and a justification for the recommendation.
2. Examine the implicit and explicit inventory policies described in the case, and evaluate them. Develop a comprehensive inventory management strategy and show how it would be applied for the two items detailed.
3. Do you think the lost customer sales should be included as a cost of inventory? How would such an inclusion impact the ordering policies you established in question 2?

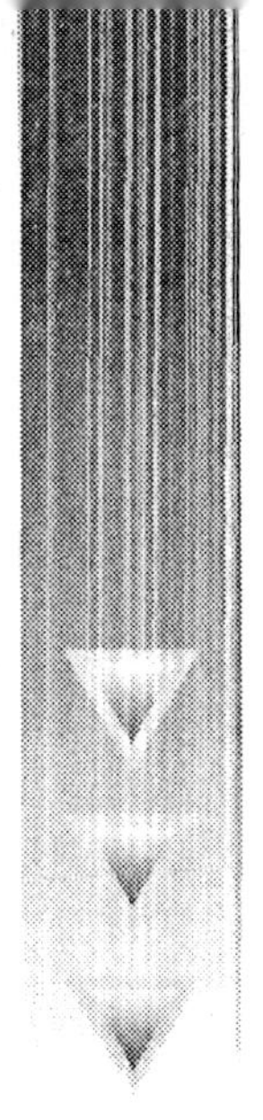

CASE 28

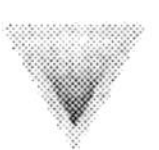

MYTOWNE FOOD STORES

MyTowne Food Stores was a small independent chain that had been in business for over 50 years. The chain had been very successful with its ability to buy in large enough quantities from producers and distribute to the individual stores through a central warehouse. Recently they were bought out by a large holding company and plans are being made to upgrade the current warehouse. The new owners are not currently interested in increasing the volume through the warehouse, but would like to see major improvements in inventory turnover, throughput time, and labor productivity with a reduction in the damage to goods during handling. The distribution center is currently running three shifts per day, five days per week with a part-time shift on the weekends. The weekend shift is very flexible in their ability to hire local college students as needed to catch up on late shipments, put away receipts from the week before, perform general cleanup duties, and ship special orders during holiday seasons. Labor costs at the distribution center have been consistently above budget with frequent overtime being used to keep up with demand. This causes disruptions when workers carry over into subsequent shifts and vie for much needed equipment. The weekend shift may even be required to go to two shifts during the upcoming holidays.

The new owners are open to productivity improvements and are willing to buy new equipment if it can be justified. Any improvements would, however, use the current facility due to its prime location to distribution channels both inbound and out.

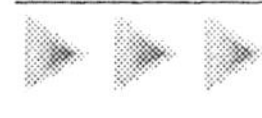

MERCHANDISE

The merchandise handled at the facility is considered dry goods, which can normally be found in any grocery store. This excludes dairy products, fresh fruit and vegetables, meat, fish, and frozen foods. Those items are handled in different facilities due to the specialized storage and shipping and handling required for each. The total volume of dry goods handled in the main warehouse is approaching 450,000 pallets per year. There are 15,000 individual SKUs in the facility. Of the goods shipped, 60% are for quantities of 1 to 10 cases. There are many small box items (30% of the SKUs) such as baby foods, specialty crackers, and low-volume imported goods that account for only 5% of the volume shipped. Some of the bulky items are identified as pallet load quantities and are transported directly to marshalling via low-lift trucks. Damage is a considerable expense accounting for approximately 1% of total goods shipped due to handling and transporting within the facility.

THE FACILITY

The current warehouse, which was originally opened in 1963, is approximately 800,000 square feet in total area (see figure). Of this, 20% consists of the original structure with usable ceiling height of 16 feet. Approximately five years ago a massive addition was made to the building, which has a usable ceiling height of 30 feet. The facility is in good shape and both the old and the new floors can withstand the weight of the new high-lift reach trucks. Offices and support services are all in the new section and consume less than 3% of the total floor space.

RECEIVING

Goods shipped to the warehouse by rail are received directly from a spur line located on the east side of the old part of the building. Goods shipped in by truck are received through 12 docks in the north end of the plant. There is a dispatching area to hold the goods received in truckload quantities, which are then transported to their home location in the picking aisles or taken to the bulk storage area. Most received goods are taken by operator-controlled tow vehicles pulling trains of up to eight carts to the bulk storage locations. Lift truck operators unload the carts into bulk storage or directly into the home locations. The bulk storage area is located on the east side of the building in the new high-ceiling section. This area is used to hold any excess goods that can't be stored in the home locations. It also allows MyTowne to obtain large quantity discounts on carload lots of canned goods and preserves in season.

ORDER PICKING

Crews of two or three operators travel through the picking aisles operating wire-guided tow vehicles that pull about five cars each. The movement of each vehicle is controlled by the lead hand for each crew via a radio transmitter worn on the belt. The other members of the crew fill the orders by picking from the home locations for each item as the train is guided down each aisle. The lead hand gives instructions, helps with the picking, and checks off the pick sheet. The lead hand also tries to keep an eye open for goods in short supply to request replenishment by the lift truck driver assigned to the area. Stopping the crews or having to go back for a few items is very time consuming and error prone. Most of the items are picked from floor-level pallet locations in the main aisles.

Some bulky items such as breakfast cereals or paper products are stacked on the floor in wide aisles for direct picking from storage. The trains travel these aisles as well. Although these products represent a small portion (20%) of the SKUs, due to their size they usually account for 50% of the volume shipped. It often takes more than one train to complete an order for the large stores. In this case the orders are computer sorted so that more than one train can be used simultaneously to pick from the main aisles.

Home Locations

Each SKU has a home location that is restocked by narrow aisle lift trucks which travel throughout their assigned aisles in cooperation with the order pickers. Overflow pallets of material are stored in the racks above the home locations. Home locations are typically at floor level and can be replenished with full pallet loads. For small volume items shelves within the pallet racking are used with up to six items stored in one location.

Lift Truck Drivers

Lift truck drivers are responsible for keeping the home locations filled and putting away newly received goods that arrive in their areas by the tow trains. They are each assigned three aisles and are very familiar with the goods in their areas. No locator system is used for goods other than for the home locations and the drivers do a very good job at randomly filling the overhead racks with goods close to their home locations. As home locations are emptied, the drivers need to also continuously remove empty pallets and place them on trains returning to receiving.

Marshalling

When the pickers have completed their respective orders the trains are driven to a marshalling area where the trains are unloaded using counterbalanced trucks, staged in a shipping lane—hopefully behind the trailer it will be shipped in—checked for completeness, and stamped with an identification for each store. There is a special cooperation between the pickers and the order checkers in that each box is placed on the carts to be visible from the outside for checking and stamping. Lift truck drivers are dispatched from the marshalling area to try to rectify shortages, which may have been due to stockouts at the time of picking. Shortages that cannot be filled are noted on the shipping documents.

SHIPPING

Once the orders are checked they are hand loaded into tractor trailers for delivery to the stores. Most of the stores require less than a truckload and individual stores' goods are loaded with last-off loaded first. Store orders are separated by nylon nets and the unloaders are supposed to use the store identification stamp to avoid mix-ups upon receipt.

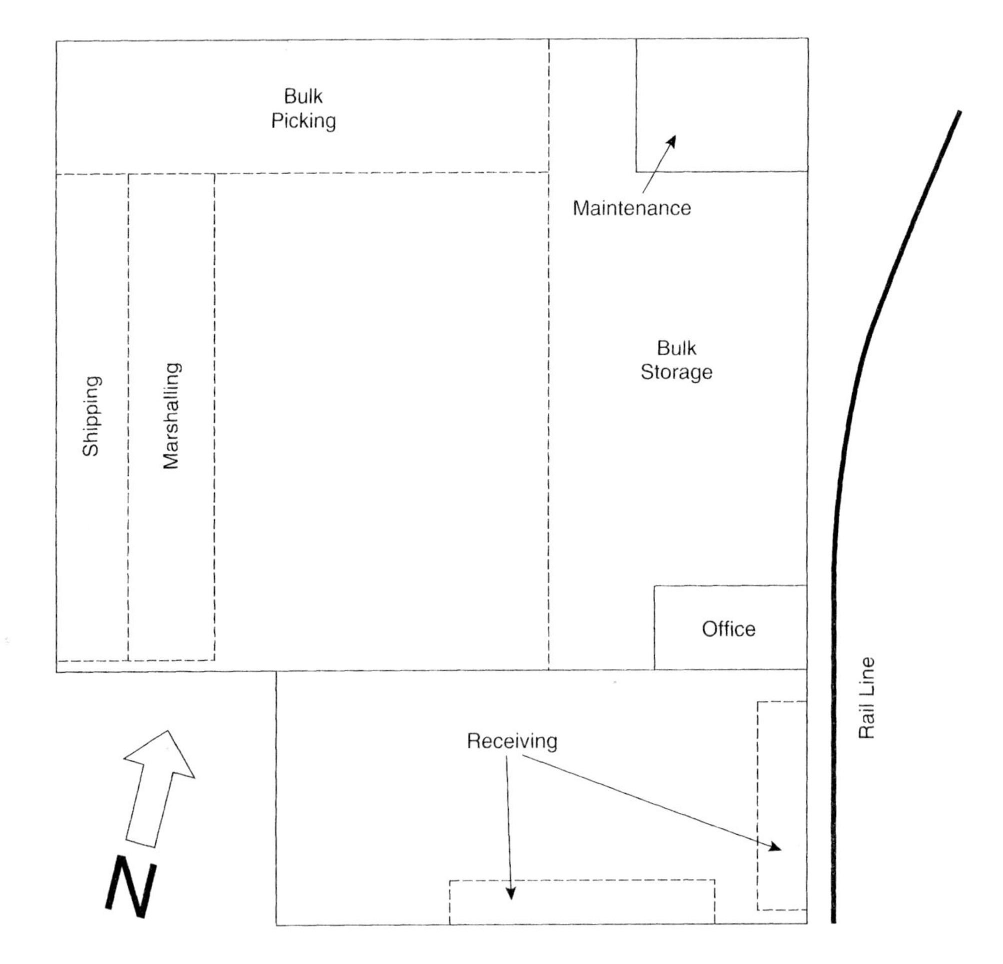

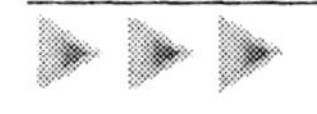

CASE ANALYSIS

1. Identify what you see as some of the problems associated with the present operations or the layout of the plant.
2. If there was enough money for modern systems such as pick-to-belt, how would you sort the goods by handling method (i.e., bulky goods, low-volume small package)?
3. What suggestions would you make about the location of the offices, maintenance, and so forth?

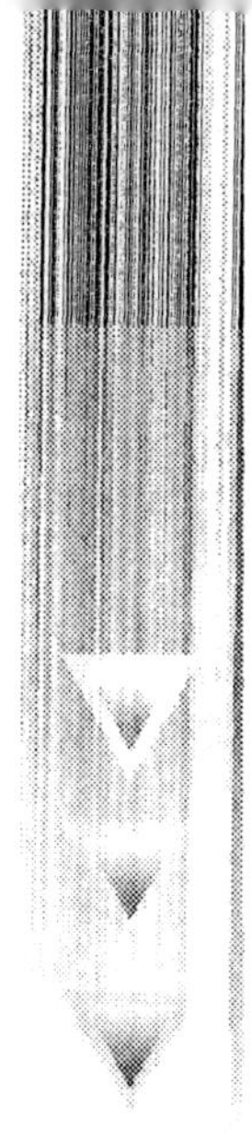

CASE 29

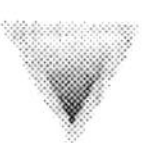

MANCHESTER MANUFACTURING

BACKGROUND

Several years ago Marcus Bremer had heard the expression "Be careful what you ask for—you might just get it," but he only just recently began to understand what it really meant. Marcus had first been employed by the Manchester Manufacturing Company as a material handler shortly after he had graduated from high school. The company was fairly new, but since it manufactured several sophisticated computer components, it had grown rapidly and was now considered a medium-size company.

After only a few months as a material handler, he was approved for an open position as a machine operator, meaning more money and more responsibility. He soon proved himself to be skillful, intelligent, and personable. It wasn't too long before he became well-liked and respected by both the other operators and the managerial staff.

A year after he became a machine operator, a production supervisor position opened up when one of the current supervisors retired. Marcus was asked to apply and subsequently was offered the job.

MANAGERIAL FRUSTRATIONS

His personality and intelligence allowed him to quickly become an effective and respected supervisor, but the position was not without some major frustrations. The major source of the frustration came from a direction he had never anticipated before he became a supervisor. A large portion of his day was spent handling material problems that constantly threatened to shut down equipment in his area of responsibility. This prevented him from spending more time working with his people to make the area run more effectively and efficiently.

Material was supposed to be automatically delivered to the production area by the material handlers according to the production schedule—a schedule that was published well in advance of need. This schedule was readily available to all functions in the operation, since it came from the modern enterprise resource planning (ERP) system recently implemented in the facility. Unfortunately, the production warehouse often delivered less material than was called for, and occasionally failed to deliver the right material at all. The reason cited was "material shortages." At other times the quantity in the container was incorrect (based on what the material tag reported as being the quantity delivered). Sometimes the material was the wrong part, even though the part number on the container said it was the correct part (in other words, the parts had been mislabeled). Marcus found the situation increasingly frustrating, and soon the frustration developed into anger.

Marcus just couldn't understand how the production warehouse could be so messed up. He had taken a short course in ERP before the system was implemented, so he knew the system should identify to purchasing the material needs well in advance—certainly with enough lead time to be able to order the right material. If that was the case, then why were things so "messed up"?

It didn't help his spirits when he received little sympathy from the other production supervisors. Most of them had been on the job for several years, and usually would just tell him, "Just get used to it—it's part of the job." Marcus just could not accept that explanation, however, so he continually complained about the situation to his boss (John Gunther), the manufacturing manager. At one point he said to his boss, "Why can't the production warehouse get their act together? I wish I had charge of that place, I'd whip them into shape in a hurry."

INTO THE WAREHOUSE

About six weeks after Marcus made the statement the warehouse supervisor quit. Mr. Gunther came to Marcus and told him, "You just got your wish. I'd like you to take over the production warehouse and put it into shape." Marcus thought he already knew some of the problems he would have to deal with, but also realized he needed to proceed carefully. He not only needed to determine if there were additional problems of which he might not have been aware from his previous production supervisor's position, but also knew he had to try to determine the source of the problems. To jump too quickly into radical solutions not only presented the danger of missing the real problems, but also increased the probability of alienating the warehouse workers. The approach he took was to conduct a fairly detailed assessment of the situation. The following is a summary of the key points he found:

- Warehouse employees could be basically categorized into two groups. The first group consisted of people hired very recently. Many appeared to be energetic with good attitudes, but were largely inexperienced. He found, however, that members of this first group would leave the warehouse often before they obtained enough experience to become competent. The warehouse workers were on the lowest pay grade in the company, and most of these new workers were really using the warehouse job as a "springboard" to move into a production operator's position as soon as one opened up in the factory. The turnover of the people in this group was very high, given the growth rate of the company and the almost constant need for production operators. Marcus approached the Human Resources manager to see if anything could be done to change the situation. The response was, "Not that I can see. The pay grades and job classifications are clearly laid out in the union contract, and the present contract has two more years to run. I don't think either side would be interested in reopening the contract before the two years is up, given the difficulty we had in getting the agreement in the first place."
- The second category of warehouse worker consisted of those who had worked in the warehouse for a very long time—several years, for most. Most of these workers appeared to lack either the capability or motivation to move into production. Many could neither read nor write, and several were being counseled for drug or alcohol addiction problems.

Marcus requested the help of the company's internal auditors to conduct an informal audit of the inventory records. The following points summarize their findings:

- The count accuracy of the inventory records was terrible. When compared to the record quantity, the real count was accurate for only 19% of the records checked—and it had only been three months earlier that the annual physical inventory had been done. The process for keeping inventory records appeared simple enough. When material was received through the receiving/inspection area (under the purchasing department), a count was entered into the computer. That count would be added to the record for the part. When parts were needed for production the warehouse worker would count out the quantity to be delivered and write in the amount on a transaction sheet. If parts were returned from production, they would be put back into storage and the quantity entered on another transaction sheet. At the end of the day these transaction sheets would all be entered into the computer by a keying operator working on the second shift.

  A physical inventory was time consuming and expensive to conduct, given that there were over 35,000 different part numbers in the warehouse.
- The location accuracy was also very bad. The warehouse used the "home base" system, in which every part number had a specific storage location assigned. The previous manager had selected this approach since every part number would then be stored in the same location at all times, making it easier to keep track of. Marcus noted, however, that the company was constantly introducing revisions of existing models, and new products were also constantly being developed. That was

very common in the computer industry, and it was not likely to change any time soon. Each new model or major revision of an existing model could mean hundreds of new components with an equal number of old components becoming obsolete. Even with the dedicated "home base" for parts, only 68% of the parts were being stored in their correct location, according to the auditor. One of the warehouse workers gave them some idea of the cause. He told Marcus, "The problem is that with all the new part numbers constantly hitting us there just isn't room on the shelves. There's plenty of open spots reserved for part numbers that we don't use anymore, so sometimes we just put things in an open spot. We intend to go back later and change the part number for the location, but things are so hectic here so much of the time that sometimes it just gets overlooked."

- Parts were constantly being returned to the warehouse from production areas because of incorrect quantities being issued or from the relatively constant changes that occurred in customer orders after production had begun. The production workers were apparently being careless about keeping the right part number with the parts, since it was suspected that many of the returned parts had the wrong part number on them. The warehouse workers often would restock the parts under the part number given by the production area, not realizing the number was incorrect. An audit of returned parts showed that almost 15% of the parts being returned were labeled incorrectly. The audit also showed that the quantity of the returns were often wrong (almost 45% of the time). The production workers were apparently also careless about checking the quantity they wrote in on the return transaction sheet.

Given the results of the audit, it was not too surprising to Marcus that he had had so much trouble with the warehouse when he had been a production supervisor. Now, however, it was his job to fix the problems. He needed to develop a comprehensive plan to correct the situation, and he needed it soon. Given his earlier boast about "whipping the warehouse into shape in a hurry," his reputation was at stake. Any thoughts about a long-term future with the company depended on a plan to turn around the situation in a hurry.

CASE ANALYSIS

1. What additional information should Marcus gather to make his turnaround plan, and why does he need the additional information you identify?
2. Develop a comprehensive plan to attack the problems. If you need to make assumptions regarding information not given in the case, state those assumptions and describe why they are reasonable given other information in the case.
3. Suppose he needed to develop a cost justification for expenses required to implement the plan. List the production costs and other expenses that could potentially be reduced by correcting the warehouse problems. In more general terms, what are the usual financial implications of having warehouse problems such as those described in the case?

CASE 30

METAL SPECIALTIES, INC.

Metal Specialties is a wholesaler of specialty metals such as stainless steels and tool steels. They purchase their stainless steel from a mill located some 200 miles away. At present the company operates its own truck. However, the truck is in need of repair and this is estimated to be about $20,000. Annual operating costs are $30,000 and the line-haul costs are $2.20 per mile. Janet Jones (J J), the traffic manager, wants to reduce the cost of bringing in the stainless steel and, because of the impending repairs, she feels now is a good time to look at alternatives. She has solicited a number of proposals and has narrowed her choices down to a motor carrier and a rail carrier.

Heavy Metal Transport (HMT), a contract motor carrier, has an excellent reputation for service and reliability. They have submitted an incremental rate, $4.00/cwt for shipments weighing less than 150 cwt, $3.80 for shipments between 150 and 200 cwt, $3.60 for shipments between 200 and 250 cwt, and $3.40 for shipments over 250 cwt up to a maximum of 400 cwt.

Midland Continental Railway has submitted a piggyback rate of $3.25 per cwt with a minimum load of 200 cwt. The piggyback rate includes pickup by truck at the steel mill, line haul by trailer on flat car, and delivery by truck to Metal Specialties' warehouse. They are considered to be a reliable carrier as well.

The finance department estimates that Metal Specialties' annual inventory carrying cost is 20%, the cost of inventory in transit is 10%, and the cost of capital is 8%. The cost of placing an order for stainless steel is estimated to be $40 per order. Stainless steel presently costs $300 per cwt.

CASE ANALYSIS

J J has to make a decision soon. Given the information above, what would you advise her to do?

CASE 31

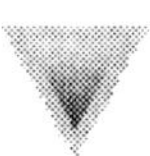

MID WEST TILE COMPANY

Mid West Tile Company, located in St. Paul, Minnesota, makes ceramic tile and fixtures used for many types of installations including bathrooms, kitchens, flooring, and wood heat installations. Sales have dropped over the past few years to 60% of traditional demand for two main reasons. Strong competition is coming from competitive products in abrasion-resistant plastics for the bath and kitchen applications and the demand for wood heat products is dropping slightly due to market saturation and a movement to more convenient fuel sources.

Mid West produces 50 standard patterns and sizes of tiles plus custom orders for large, commercial installations in hospitals, hotels, institutions, and factories. The major raw material for the tiles is locally mined clay. However, many of Mid-West's products are blended with specialty clays imported from a geographically unique deposit in Brantford, Ontario. The amount of the special clay used in each product depends on the need for certain characteristics such as heat resistance, finish, color, and strength. Demand for the specialty clay is estimated to be 90 tons per year for the foreseeable future.

There have been some changes in management, and the company is looking for all possible ways to reduce the costs of goods sold and improve cash flow. Betty Mansfield, the raw material buyer, is concerned with the total cost of sourcing and storing the clay. She has contacted the supplier to learn the costs at various points in the supply chain and at various quantities.

The specialty clay is imported into the United States duty-free and varies in price according to the amount ordered at one time. Currently the company orders in

50-ton (carload) quantities to minimize cost, however, Betty has questioned the need to order shipments only twice a year. She has contacted the supplier who faxed her the following costs:

Quantity	Method of Shipment	Cost
50 tons	Rail	$37.60/cwt
35 tons	Rail	$38.40/cwt
20 tons	Road	$39.10/cwt

The above costs are FOB* St. Paul, and the shipments take an average of two weeks by rail and one week by road.

Carrying costs of the clay are estimated to be 20% per year. This includes the cost of capital plus costs associated with storing the clay. Although the clay is a natural product, which in its raw form in the ground is exposed to moisture and a wide range of temperatures, once blended the clay must be kept from freezing and exposure to moisture to prevent clumping and color variations in the final product. The clay is shipped in 80-pound bags, which is a standard size for the supplier. The individual bags are also easy to handle in the factory and facilitate blending and measuring.

The variability of rail shipments has been a problem in the past, since the product is routed via Toronto, Chicago, and finally St. Paul. This can lead to a lead time of up to four weeks. As a result safety stocks for rail shipments are set at one month of average demand. Road shipments are very consistent in their timing. Severe winter weather does occur, however it is very rare for a storm to delay shipments by more than two days.

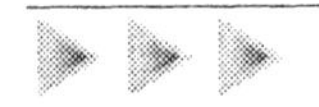

CASE ANALYSIS

1. How much safety stock should be kept if the clay is shipped by road?
2. Calculate which order quantity would be used to minimize the total yearly cost of buying and storing the clay.
3. The supplier has just left a message with Betty stating that the clay could be supplied at a cost in Canadian dollars of $50.25 per cwt FOB Brantford.** Would this be a good way to source the clay? What else should be taken into consideration with this method of sourcing? (You will need to find your own information on Canadian-U.S. exchange rates, etc.)

Inco Terms
*CIF
**Exworks

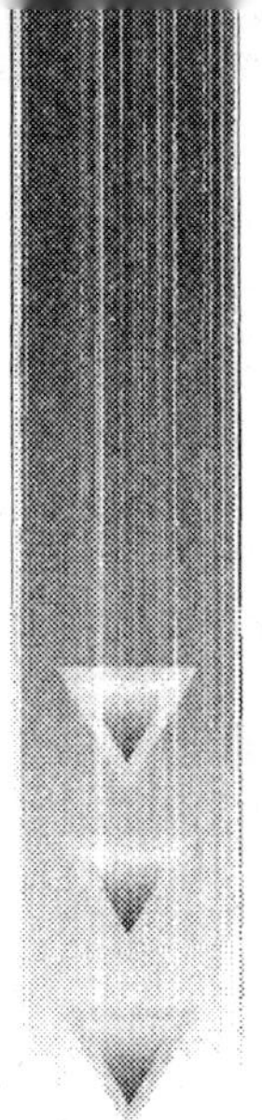

CASE 32

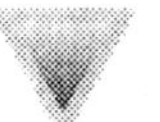

SKIMMER WATERCRAFT

Skimmer Watercraft located in Springfield, Illinois, manufactures a variety of boats for the recreational market. The boats are made from fiberglass cloth, resin, wood, and metal. Skimmer also supplies their product with engines that are installed complete and warrantied separately from the boat itself. Sales to independent distributors are fairly constant throughout the year with discounts used to encourage sales in otherwise slow periods. The plant runs continuously with few interruptions, even for model changes.

An analysis of freight volume showed that 80% of the weight shipped (in ton-miles) was carried by 12 carriers. These goods came from Ontario, New York, Kentucky, Pennsylvania, North and South Carolina, Florida, California, and Texas. Resin, wood, engines, and glass cloth made up the majority of the weight shipped. These same 12 carriers also handled the medium volume materials such as hardware, windshields, steering wheels, and upholstery, which are purchased from a number of smaller suppliers. The remainder of the freight was made up of short-haul local freight and many small courier shipments from a large number of specialty suppliers.

Skimmer is having trouble controlling inbound shipments. Each supplier is responsible for arrangement and payment of the freight and they usually ship LTL. An order point system is used to indicate the need to replenish raw materials, which results in a somewhat irregular inbound shipment schedule. All of the suppliers were very reliable. However, in-transit time for shipments was quite varied since the goods traveled LTL. Transit times varied from three to 12 days depending on the distance traveled and the routing of other goods in each load.

Occasionally Skimmer ran out of raw materials for the plant, which would disrupt production schedules and threaten on-time delivery of customer orders. Purchasing is constantly trying to locate shipments and asking for expedited service including air where necessary. Inventory levels are higher than planned, since there is a tendency to order goods early in an attempt to overcome the variability of the lead times. All of this firefighting is disruptive to the purchasing staff who don't have time to perform their regular tasks. A request was made to increase staff, but management would prefer to see the problem fixed rather than simply increase indirect costs.

The receiving and warehousing staff are also under a lot of pressure. The inbound traffic is sometimes very heavy with many shipments arriving at once followed by periods of inactivity between shipments. Suppliers were becoming impatient with Skimmer because their trucks were often delayed due to congestion at the loading dock; at times making the drivers park on the street. This has even happened with expedited shipments that the carriers had gone out of their way to get to Skimmer on time only to be delayed at the receiving dock due to apparently poor traffic management. Drivers were now requiring overtime and other deliveries were being delayed.

Analysis showed that inbound shipments varied between 100 and 25,000 pounds with the average being about 6,000 pounds per shipment. On a monthly basis eight common carriers hauled approximately 400,000 pounds of goods from the East and Ontario. Four common carriers were used to ship 200,000 pounds per month from the West. During an average month there were in excess of 240 deliveries made of which about half were small local carriers or small package couriers.

CASE ANALYSIS

It was obvious that something had to be done about this situation. It was costing Skimmer a great deal in production, warehousing, and carrier problems. What do you think they should do?

1. How would you organize the inbound freight to reduce the number of carriers?
2. List the advantages to reducing the number of carriers.
3. Would outbound freight influence your choice of carriers?

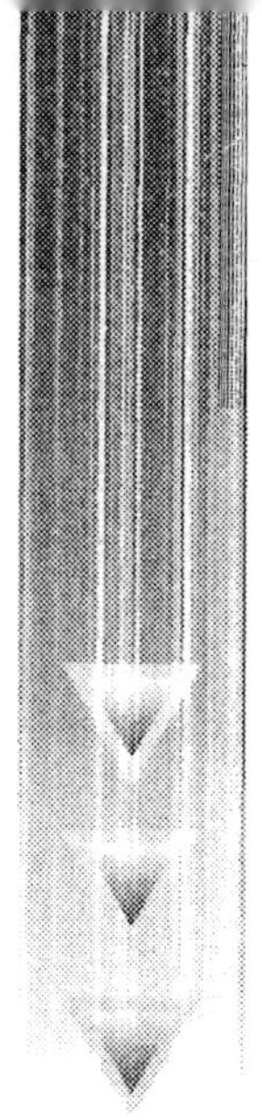

Case 33

Bremer Packaging

Bremer Packaging of Syracuse, New York, manufactures custom packaging for the food and beverage industry, and has also made some inroads into protective barriers for medical products. The finished products use various combinations of paper, polyethylene, and foil to protect products from dirt and moisture in the environment, as well as to contain liquid, dry, or solid products. The Syracuse plant does not manufacture the laminate but converts the large web roll to custom products by cutting to shape or width and printing directly on the laminate. Some familiar examples of their consumer market products include peelable foil enclosures for individual serving dessert products, high resolution graphic seals for restaurant serving salad dressings, and blister packages for small hardware items. All their products are made from films imported in "jumbo" form from parent company facilities in Bremer, Germany.

The jumbos are very large rolls (approximately 40 inches in diameter) that may weigh in excess of 1 ton depending on the product. The product is shipped on a regular basis from Bremer where it is stowed into 20-foot ISO ocean containers. Each container has a capacity of only 14 rolls, since the rolls are too heavy to stack without crushing the edges of each roll. The rolls are protected with an outer wrap to prevent damage during handling while stuffing the container as well as damage from impacts during movement of the container. Only a small amount of dunnage is required for the last row of rolls to secure the entire load.

The containers are stuffed at one of the German handling facilities in Bremer and then sent to a container port and loaded on to a ship bound for the deepwater port of New Jersey. Upon arrival, the container is processed through U.S. Customs and sent by road transport to the plant in Syracuse. The entire journey can be accomplished in 15 days, but for planning purposes, a 20-day lead time is used in case of delays due to weather or problems with Customs documentation. The planned lead has proven quite reasonable with most shipments actually taking 18 days from dispatch to receipt in Syracuse. The cost of transportation for each 20-foot container shipment is approximately $2,000.

Inventory policies for each of the various packaging films use a safety stock of one month. This amount was determined somewhat arbitrarily, but it was felt that overforecast demands for an individual product or rejection of some material due to shipping damage could be rectified by ordering an extra shipment, which would be available within the month of protection.

The excessive costs of less-than-container-load shipments dictate the order quantity for each product to be one container load or 14 rolls of packaging film. The rolls of film vary in value, but for inventory purposes are valued at $4,200 each.

The current inventory records for their top 10 products are as follows (all units are in full jumbo rolls):

Item #	Product Group	Annual Usage	On-Hand	On Order
1056	Rhino Foil	42	5	14
1057	Rhino Foil	211	31	0
1058	Rhino Foil	78	19	0
1061	Rhino Foil	31	16	0
1142	Tough-Hide	49	13	0
1143	Tough-Hide	18	12	0
1145	Tough-Hide	47	8	0
1221	Clear-Seal	39	3	14
1225	Clear-Seal	33	9	0
1967	E-Z Peel	97	11	14
1977	E-Z Peel	59	14	0

Bremer has begun investigation of an ERP system, however as the materials manager you feel that some of the concepts of just-in-time and balanced flow could be applied to the shipments of packaging film. To start your investigation and lay the groundwork for any justification of expenditures, conduct the following analysis.

CASE ANALYSIS

1. Calculate the theoretical and average inventory, the actual inventory, and the in-transit inventory.
2. Apply the just-in-time principles to shipments and describe the changes you would make to ordering and shipping. Calculate the new levels in question 1.
3. Assume a reasonable carrying cost for inventory and calculate annual savings.

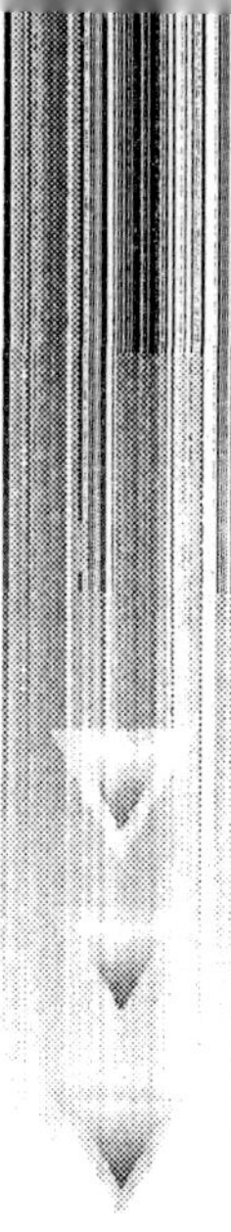

CASE 34

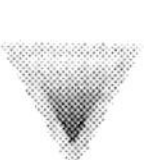

MURPHY MANUFACTURING

When Joe Vollbrach, Vice-President of Operations for Murphy Manufacturing, was given the CEO's directive in 1981 to investigate the new Just-In-Time (JIT) concepts and to implement them if appropriate, he was slightly apprehensive. Everyone knew, he thought, that MRP was the best way to run a manufacturing operation, and they had been pretty successful with their MRP system. Once he read a couple of books and a magazine article or two about JIT, however, he thought maybe there was something to it—and it sure seemed simple enough. Dozens of companies had reported great reductions in inventory cost, and with Murphy Manufacturing having only between five and six inventory turns per year, the prospect of significant inventory reductions appeared very appealing.

Encouraged with the success stories and very mindful of the CEO's directive, Joe wasted no time. He put out the directive to all his people to implement JIT the way it was working in the book examples he had read. Early in 1982, however, he was beginning to wonder about the truth of the success stories in those books. The following are some of the examples of the complaints he was getting and the problems he was facing.

Karen, the Purchasing Manager, "Joe, this JIT is a disaster for us. It's not only costing us a lot more money, but the suppliers are getting real angry with us. Since our raw material inventory had typically been high, you said we should order smaller quantities and have it delivered just in time for its need. Sure, that cut down on the raw material inventory, but that cost saving has been more than made up for with all the increased cost. First, purchase orders are not cost free and we're making a lot more of them. That's also taking up a lot more of our buyer's time.

Then there's the transportation cost. Since most of the trucking companies charge a lot more for less-than-full truckloads, our costs are going sky-high with more frequent deliveries of smaller loads. Combine that with expediting costs, and it gets really bad. Our schedules are changing even more frequently, and without the raw material the production people are often asking for next-day delivery of material they need for a schedule change. We're flying more parts in, and you know how much that costs!

"That's not all, our suppliers are really wondering if we know how to run our business. We're changing the schedule for them much more frequently, and the only way they can hope to meet our needs is to keep a lot of our inventory in their finished goods. That's costing them a lot of money in inventory-holding costs as well as administrative costs to manage the inventory and to process all our requests. They not only have more requests from us, but it seems like everything is a rush order. They're pressing us hard for price increases to cover their increased cost to keep us as a customer. I've held most of them off for a while, but not much longer. Unfortunately, I agree with them, so it's hard for me to make any kind of logical argument to counter their requests for price increases."

Oscar, Supervisor of Shipping/Receiving, "If you're going to keep this up I have to ask for two more truck bays and about four more receiving clerks. There's a lot more trucks making a lot more deliveries. We can't schedule perfectly when a truck will show up so many times during the day that there are several trucks waiting for an open truck bay. Production people are screaming at us because they see a certain truck and they know a critical part is on it, but we have to wait because the trucks currently in the bays being unloaded also have critical parts. It seems like everything is critical these days. A few more bays with enough people to staff them will help, although it's not the only answer. I figure about a $2 million renovation to the receiving area and an increase in our budget of about $300,000 ought to be about enough to keep us afloat, at least most of the time. Oh, I forgot, better make that $350,000 more. I also need another records clerk to keep up with the big increase in paperwork from all those deliveries."

Joe had just finished taking a couple of aspirin to deal with his headache when Marsha, a production supervisor, caught up with him.

"Joe, I don't know where you came up with these silly ideas, but you're killing us here. This so-called "Kanban" is just silly. It reminds me of the reorder points we used before the MRP system, and we all know how much better MRP made things work. I feel like I just slipped 25 years back in time. We all know our customers are a fickle bunch who are always changing their minds on orders. We used to have visibility with the change when we used MRP, but now all we have is what the final assembly area is working on. That not only causes us big, inefficient downtimes for setup changes, but also doesn't give us much time to get the materials we need to make the different parts. When we do figure out what we need, most of the time they don't have any in raw material stock. Now we have to waste more time and energy to scream at the buyers. What makes it even worse is your directive to eliminate finished goods. We don't have material to make what we need half the time, and we're not allowed to make what we can. It would seem to make sense that when a customer does request a certain model we have in stock, at least that order wouldn't end up being a crisis."

"By the way, some of our best people are threatening to quit. Some are being paid on a piece rate, and the lack of material combined with the directive to avoid finished goods has cut seriously into their paychecks. Even the ones on straight hourly pay are unhappy. All the supervisors are being evaluated on labor productivity and efficiency, and our numbers are looking really bad. Naturally, we put the heat on the workers to do better, but I'm not too sure they can do much about it most of the time."

"And another thing while I have your attention, in some of our models we have a lot of engineering design changes. The MRP system used to give us some warning, but now we have none. Suddenly one of the engineers will show up and tell us to use a different part. When we check with purchasing, often they've only just gotten the notification themselves and have only just started to work with the suppliers for the part. Not only do we have to obsolete all the old parts, but also we have to wait for the new ones. The salespeople are even starting to scream at us."

"While I'm on a roll here, you've got to do something with those quality control people. We could deal with some scrap or rejected parts when we had plenty of inventory, but now we need every part in the place. Tell them to quit rejecting parts and let us use them. At least we'll have a better chance of shipping the order, even if it isn't perfect."

Almost as soon as Marsha left, Valerie (the sales manager) came in. Joe braced himself for more of a headache even before she spoke. The expression on her face foreshadowed what she had on her mind.

"Joe, my job is to make sales and keep the customers happy. The sales have been going fine, but our delivery stinks. Our on-time delivery record has fallen from 95% to less than 50% in the past six months. Some of our customers are threatening to leave us, and some are pushing for price cuts. According to them, our delivery record is so poor they feel compelled to keep more of our inventory in their raw material stores to account for their lack of faith in our delivery promises. They say that since it's our fault they have increased raw material inventory costs, we should compensate them with a price cut. It's pretty hard to argue with their logic. I'm sure we would do the same with our suppliers if they treated us the way we've been treating our customers lately.

"We all know our customers have to sometimes change orders to reflect what their own customers want. Now, however, the changes are becoming more frequent and radical. It seems that since we are so poor in delivery they order more in advance from us to buffer the time for our late deliveries. Ordering further into the future gives them a lot less certainty of what they really need, so naturally they have to change once they do know. I may have to get another order entry clerk to deal with all the changes, and if I do you had better believe I'll let everyone know the extra expense is your fault!

"The bottom line here is simple, Joe. You've got to get after your people to improve the delivery drastically or we may be in big trouble, and soon."

After Valerie left Joe's secretary came in with a rush memo from the CEO.

"I just got the preliminary financial report for the fourth quarter of 1981, and for the first time in over five years we show a loss—and it's a big one. The details show sales goals have basically been met. The loss comes from a very large increase in expenses in virtually every area in operations except a modest decrease in inventory cost. I'm calling an emergency staff meeting for two o'clock today. Please be ready to explain the situation completely."

As Joe shut his door to insulate himself from the complaints and to prepare for the meeting, he began to wonder if some of the people who claimed that JIT was a culturally based system impossible to implement outside of Japan were correct. He knew most of the problems were related to JIT. What, he wondered, went wrong and what should he do about it? These were certainly two critical questions that would be major parts of the two o'clock meeting and he needed good and complete answers. The CEO was reasonable and could deal with the fact that mistakes might have been made, but she would expect a detailed analysis and complete action plan to get things back on track. The aspirin were definitely wearing off and it was less than an hour ago that he took them!

CASE ANALYSIS

Prepare a complete report for Joe to use for his two o'clock meeting. This should include both the analysis of what went wrong and why, as well as a comprehensive and time-phased plan to implement JIT the right way. If you do not think JIT is appropriate, explain why in detail and develop a comprehensive alternative plan.

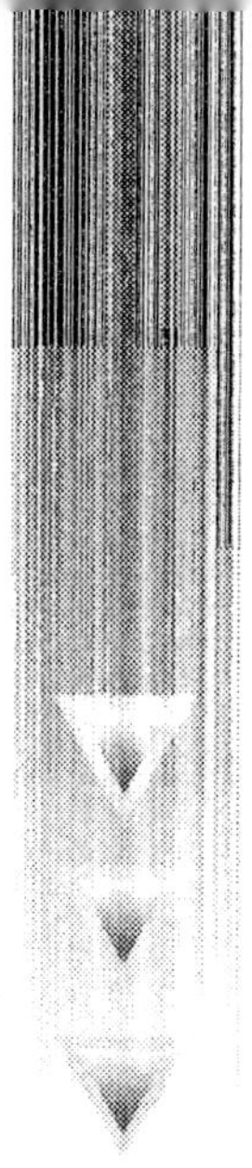

CASE 35

ACCENT OAK FURNITURE COMPANY

Accent Oak Furniture Company has been in business for more than 30 years serving customers in Chicago and the surrounding area. Their business consists of three different divisions composed of a bannister division that manufactures and installs quality oak railings, a custom furniture division that manufactures such crafted items as dining room suites, and a kitchen cabinet division.

Each division reports to a vice-president who in turn reports to Frank Johnson, the president and founder of the company. Total sales for all three divisions are projected to reach $8 million for the current fiscal year.

Frank has been quite pleased with the performance of the custom furniture division and the kitchen cabinet division, which together account for more than 85% of the sales revenue. He does, however, have some concerns about the bannister division. This concern is based on last year's profit performance and the first week of August's sales report for the installation crews (see Figure 1).

It is now the second week of August and the new installation market is just starting up. Bannisters are installed in new homes in the fall when the homes are nearly finished and ready for sale. There is some demand for bannisters throughout the year for renovations. However, this business is also concentrated in the fall and quickly dies off before Christmas.

On Monday, Frank met with Tom Smythe, the vice-president of the bannister division, and voiced his concerns. Frank suggested that they meet again on Friday to discuss some further actions to get back on track. Tom felt the concerns were a bit premature but agreed grudgingly. Tom felt that some of his key people should attend so Hank Strong, the manufacturing supervisor, Brian Coulter, the sales manager, and Pete Harburg, who was in charge of the five 2-man installation crews were invited to attend.

During the meeting, Tom asked each one of his people to describe their individual concerns about the past week's installation report.

Hank began by giving a brief outline of the manufacturing process that he was responsible for in the plant.

Handrail Area

1. The oak was purchased in 16-foot lengths and inspected for any flaws or excessive knots to ensure it was #1 grade.
2. Any rejects were sent to the cabinet division to be used for interior shelves or to be used to make spindles.
3. The approved boards were run through a multihead mill that planed the bottom side and shaped the top and sides of the handrail.
4. They were inspected again for rough grain and any knots that could cause quality problems. Any rejects were then cut into shorter usable lengths for shorter handrails or spindles.
5. The handrails are then sent to the drilling machine where the holes are automatically spaced and drilled by the operator on an industrial quality drill press.

   Occasionally the operator inspects the dimensions of the holes using a dual purpose GO-NO-GO gauge. On one end of the gauge there is a metal rod that is inserted in the hole to measure the depth. The rod has a hatchmarked area that has PASS stamped into the metal. If the top of the hole is in the PASS area then the hole dimension is considered okay. The other end of the gauge is long and round and has three steps cut into the circumference. This makes the rod get bigger around at about 1/4 inch from the end and 1/2 inch from the end. If the first part goes into the hole but not the middle, the hole is undersized. If the second step but not the third enters the hole, the hole is of the correct diameter. The third step indicates oversize.
6. The handrails are then sent for sanding and varnishing. A final inspection is performed for appearance and finish. The handrail sections are then sorted by length for full size and shorts.

Spindle Area

1. Oak is purchased, inspected, and cut to length. Approximately 5% also comes from the handrail area.
2. Sections are placed on a planer and run through twice to give the square end dimensions. Pieces are then placed into a pattern lathe, which provides the product with the spindle design and tapered top end. The operator also lightly sands the spindle while it is on the lathe.

3. An occasional inspection is performed using a GO-NO-GO gauge. The inspection procedure requires the operator to fit over the end of a spindle a gauge, which is an aluminum bar with a specified hole size. The spindle will pass inspection if 1/4 to 1/2 inch of spindle protrudes through the gauge. The gauge is actually marked with red paint outside the limits to assist the operator when inspecting.
4. The operator sets the preangled cutter to maintain the desired size of the spindle end.
5. The approved spindles are then sent to final sanding and varnishing and final inspection.

   Hank felt that his people did the best job possible given the wood and machinery available.

Brian Coulter, the sales manager, felt that his people were certainly doing their jobs since sales had increased at a rate of 15% over the last three years. One major concern that Brian did have was the increased number of complaints from his major customer, Lincoln Homes, for the quality of the installations. They have already told Brian that they will take their business elsewhere if the quality does not improve immediately.

Pete was the most vocal of the group, since his area was always getting the flack. He felt his crews were being pushed as hard as possible but just couldn't keep on plan. Accent Oak bannisters are designed to be easily installed in the field even for custom work. However, his crews seem to have to refit every other piece and spend their time chasing defects. He recently started having them keep track of reasons why they were over or under budget on each job.

Frank angrily told his people that he didn't care whose fault the problems were, he just wanted them fixed yesterday and stomped out of the room.

Tom and his group had just recently completed a Quality Management course and all agreed that they might as well see if it could help them solve the problem and get Frank off their backs.

Figure 1 Installation Department Monthly Report—August 8.

Job#	Crew#	Budget $	Actual $	Comments
7156	1	1,100	1,127	Waited 1/2 hr for customer
7157	2	985	1,154	Fit problem with spindles
7158	5	1,200	996	
7160	4	1,500	1,854	Recalled for loose spindle
7163	2	850	865	Two split spindles
7166	5	1,200	1,385	Fit problem with spindles
7167	1	1,450	1,620	Customer changed design refit
7168	4	1,800	2,254	Spindle shims needed

(continued)

Figure 1 Installation Department Monthly Report—August 8.

Job#	Crew#	Budget $	Actual $	Comments
7169	5	1,100	1,080	
7171	2	980	1,200	Handrail rough finish
7172	4	1,560	1,860	Loose spindles
7174	1	1,200	1,650	Not to drawings
7175	2	975	1,320	Handrail cracked
7177	4	1,400	1,875	Fit problem spindles
7179	3	2,250	3,200	Recalled for loose bannister
7181	5	1,900	2,520	Fit problem spindles
7182	3	1,800	2,260	Fit problem spindles
7184	3	1,750	1,780	Customer changed design
TOTAL		$25,000	$30,000	

INSPECTION REPORT—SPINDLE

Figure 2 below shows the specified depth and allowable tolerance for the pilot study on spindle fit using the knife-edge gauge designed for this purpose. The following 50 readings have been taken and recorded as of August 14.

Figure 2

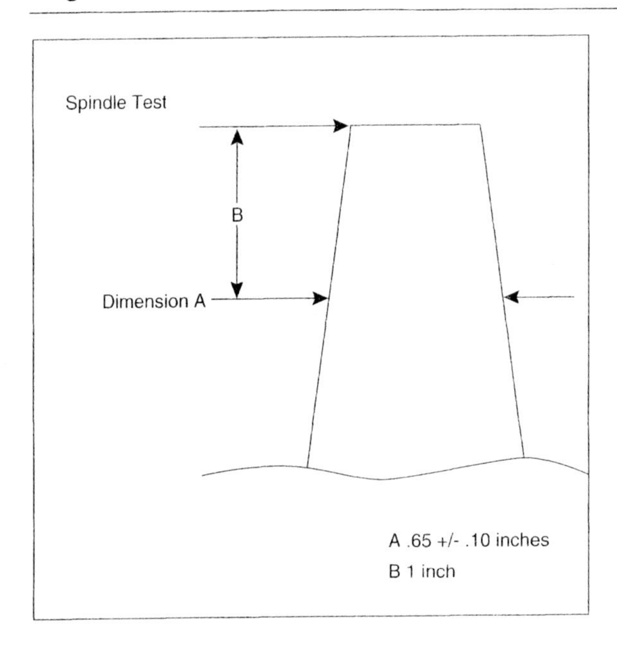

Figure 3

Spindle Widths - Inches									
.60	.63	.60	.62	.60	.59	.61	.67	.57	.61
.62	.59	.61	.64	.67	.66	.69	.63	.69	.68
.61	.67	.68	.67	.60	.61	.68	.60	.62	.60
.66	.60	.63	.62	.68	.67	.62	.70	.67	.68
.58	.68	.67	.69	.58	.69	.65	.68	.59	.64

Figure 4

Hole Diameters - Inches									
.57	.56	.58	.59	.57	.58	.56	.58	.56	.58
.58	.59	.60	.55	.61	.60	.58	.57	.58	.60
.56	.61	.57	.59	.58	.57	.55	.57	.59	.57
.60	.58	.56	.60	.56	.62	.59	.58	.62	.59
.57	.62	.59	.61	.63	.59	.64	.60	.61	.63

INTEROFFICE MEMO

TO: Lead Hand—Handrail DATE: August 17

FROM: Hank Strong

SUBJECT: Customer Complaints—Bannisters

INTEROFFICE MEMO

TO: Lead Hand—Spindle Department DATE: August 17

FROM: Hank Strong

SUBJECT: Customer Complaints—Bannisters

INTEROFFICE MEMO

TO: Tom Smythe DATE: August 17

FROM: Hank Strong

SUBJECT: Customer Complaints—Bannisters

CASE ANALYSIS

1. Analyze the customer complaints using Pareto Analysis to identify the largest cause of complaints.
2. Construct histograms of the spindle and hole data to see if there is assignable cause to the problem.
3. Complete the 3 office memos with your recommendations.